THE COMPUTATIONAL STRUCTURE OF LIFE CYCLE ASSESSMENT

ECO-EFFICIENCY IN INDUSTRY AND SCIENCE

VOLUME 11

The titles published in this series are listed at the end of this volume.

The Computational Structure of Life Cycle Assessment

by

Reinout Heijungs
Centre of Environmental Science,
Leiden University, Leiden, The Netherlands

and

Sangwon Suh
Centre of Environmental Science,
Leiden University, Leiden, The Netherlands

KLUWER ACADEMIC PUBLISHERS
DORDRECHT / BOSTON / LONDON

A C.I.P. Catalogue record for this book is available from the Library of Congress.

ISBN 1-4020-0672-1

Published by Kluwer Academic Publishers,
P.O. Box 17, 3300 AA Dordrecht, The Netherlands.

Sold and distributed in North, Central and South America
by Kluwer Academic Publishers,
101 Philip Drive, Norwell, MA 02061, U.S.A.

In all other countries, sold and distributed
by Kluwer Academic Publishers,
P.O. Box 322, 3300 AH Dordrecht, The Netherlands.

Printed on acid-free paper

Printed in the Netherlands.

Preface

This books presents a complete overview of the computational aspects of life cycle assessment (LCA). Many books and articles have been written on LCA, including theoretical treatments of the entire concept, practical guidebooks to apply the technique, and concrete case studies in which LCA is applied to support decision-making with respect to environmental aspects of product alternatives. However, a good discussion of the computational structure of LCA is lacking. Knowledge is only partially documented, and what is documented is fragmented over diverse publications with mutual inconsistencies in approach, terminology and notation.

The book is the result of several years of research, along with the teaching of LCA at university classes and, not unimportantly, the development of software for LCA. This software has been designed to support the education of LCA, but it has been applied in real-world case studies as well. The name of the software is CMLCA, which is an abbreviation of Chain Management by Life Cycle Assessment. This program can easily be used to reanalyse and further explore the ideas that are outlined in this book. Another important source for this book relates to the work involved in connecting input-output analysis (IOA) to LCA. Software for this – MIET, an abbreviation of Missing Inventory Estimation Tool – is also available. Some of the basic routines have been implemented in Matlab script as well. All three pieces of software can be accessed, free of charge, through http://www.leidenuniv.nl/cml/ssp/software.html.

In developing the ideas that are written in this book, we have benefited from discussions during the last few years with Jeroen Guinée, Gjalt Huppes, René Kleijn and Ruben Huele at the Centre of Environmental Science, Leiden University, Rolf Frischknecht at ESU-services, ETH Zürich, Mark Huijbregts, formerly at the Interfaculty Department of Environmental Science, University of Amsterdam, now at the Department of Environmental Science, Nijmegen University and Wang Hongtao at Sichuan University. Igor Nikolić provided support in discovering the advanced features of typesetting with LaTeX. The actual text, including possible omissions and errors, however, is our responsibility.

Contents

Chapter 1

Introduction

This chapter introduces the aim of this book and motivates the importance of its topic. It does so in relation to a brief introduction of life cycle assessment (LCA), in which the various types of activities are outlined as well. Finally, the structure of the book is presented, along with a reading guide.

1.1 Purpose of the book

1.1.1 Aim

This book presents and discusses the computational structure of life cycle assessment. Under the computational structure, we will capture the arithmetical rules that are involved in carrying out an LCA study. However, this book is not a book with computational recipes only. Two other aspects receive a large emphasis as well. These are the background of the computational recipes, including argumentations and proofs, even though sometimes heuristically, references to related mathematical rules, and aspects that relate to the numerical implementation of the computational recipes. For this latter, the book will not provide computer source codes, but it will concentrate on the algorithmic aspects, even though some example pieces of Matlab code are given in Appendix C. Thus the computational structure is understood here to cover the mathematical structure as well as the algorithmic structure.

The computational structure will be formulated in terms of explicit mathematical equations. It will become apparent that use of matrix algebra provides an elegant, concise and powerful formalism. One should note that the term 'matrix' in this book refers to a rigid mathematical concept (see Appendix A), that is defined in a linear space and for which operations such as multiplication, transposition and inversion are defined. Thus, Graedel's (1998, p.100) concept of matrix as a table of 5×5 cells in which the user is supposed to enter an ordinal score between 0 ("highest impact") and 4 ("lowest impact") is outside the scope of the present book.

It will be assumed that the reader has a basic knowledge of the principles, framework and terminology of LCA. Useful texts at varying levels of depth are provided by Lindfors *et al.* (1995), Curran (1996), Weidema (1997), Jensen *et al.* (1997), Hauschild & Wenzel (1998), Wenzel *et al.* (1998), UNEP (1999), Guinée *et al.* (2002), and others. However, a short overview of the basic elements of LCA is discussed in the next section. We also will, as much as reasonably possible, adhere to the ISO-standards for LCA (ISO, 1997, 1998, 2000). At certain points, departures will be necessary, and at many places, new concepts must be introduced. When appropriate, such cases will be argued.

Throughout this book, it will be assumed that data availability is not a problem. In fact, the efforts and measurement, modeling and estimation techniques that are needed to obtain data is not discussed in this book. The central theme is how the data, once available, should be processed and combined to complete an LCA study. In the first few chapters, it will moreover be assumed that data are known exactly. This will allow us to present the basic structure in terms of deterministic equations. Chapter 6 discusses extensively the topic of perturbation theory, which includes the statistical processing of stochastic data.

1.1.2   Motivation

The main motivation for writing this book is that the computational structure is an important topic for which no reference book is available. Below, we first seek to explain that indeed the topic is underemphasised, and then will demonstrate its importance.

It is a remarkable fact that there is a large number of guidebooks for applying the LCA technique, but that the computational structure of LCA is hardly addressed in these books. To some extent, this is understandable: a person charged with carrying out an LCA study needs guidelines on which data to collect, which choices to make, and how to report assumptions and results. For the calculations, he or she will rely on LCA software, of which there is a large choice on the market (Siegenthaler *et al.*, 1997). But this alleged lack of direct utility is not a decisive argument, since most guidebooks on LCA discuss the backgrounds of, say, models for ecotoxicity, even though these models are not used in an LCA, because it is only the tabulated characterisation factors that are derived from such models that are used. So, lack of direct utility when executing an LCA is not a valid reason for excluding material on the computational structure in guidebooks for LCA.

A further remarkable fact is that the computational structure is by and large overlooked by the theoretical literature on LCA as well. The equation which forms the basis for almost the entire book is

$$\mathbf{s} = \mathbf{A}^{-1}\mathbf{f} \tag{1.1}$$

in which $\mathbf{f}$ is the final demand vector, $\mathbf{A}$ is the technology matrix (and $\mathbf{A}^{-1}$ its inverse), and $\mathbf{s}$ is the scaling vector; see Sections 2.1 and 2.2 for a full explanation. In the standard literature on LCA, this equation, as well as the terms final demand vector, technology matrix and scaling vector are missing entirely. And the few sources in which the computational structure is discussed are used in a rather limited way. An example may illustrate this. In 1994, one of the authors published a paper (Heijungs, 1994) that explicitly discussed some important elements of the computational structure of LCA. It introduced a matrix formalism towards the inventory analysis, and it gave a small example system with only four unit processes with a feedback loop that needed a matrix approach for a reliable solution. Six years later, in 2000, virtually all commercially available LCA programs were still unable to reproduce these results. Some of the programs refused to perform the calculation, others gave a totally wrong answer, and still others gave results that at best approximated the exact solution.

One might think that the computational structure of LCA is a too obvious issue to discuss in scientific publications. This is suggested by the formulation in the ISO-standard for inventory analysis: "Based on the flow chart and system boundaries, unit processes are interconnected to allow calculations on the complete system. This is accomplished by normalising the flows of all unit processes in the system to the functional unit. The calculation should result in all system input and output data being referenced to the functional unit." (ISO (1998, p.10)). The forerunner of the ISO-standard, SETAC's Code of Practice (Consoli *et al.* (1993)), provides some more information, but is still far from being exact and operational on that topic. Fecker (1992, p.4) writes in a book with the promising title *How to calculate an ecological balance?* that "the process parameters are multiplied with the corresponding factor by which the process participates in the system." In this, he is one of the few authors that explicitly introduce the concept of scaling factors, but he does not provide a method to obtain them in a concrete situation. The report of SETAC's Working Group on Inventory Enhancement (Clift *et al.* (1998)) ignores the topic entirely. Another famous SETAC-publication (Fava *et al.* (1991, p.15)) is more explicit: "The calculation procedure is relatively straightforward ... The calculations can usually be performed by common spreadsheet software on a personal computer." This is, however, no longer true. As we will see in subsequent chapters, the theory involves concepts such as linear spaces, singular value decomposition, the pseudoinverse of a matrix, and the condition number of a matrix. Of course, there are a few texts in which the topic is addressed. For an overview, see Section 1.3.

It is the authors' experience that a good knowledge of the computational structure of LCA is important for several reasons:

- it is a prerequisite in the construction of a method that really can claim to have scientific validity;

- it is useful to gain an understanding of the logic of LCA in a university course;

- it guides the design and implementation of reliable LCA software (so proves the aforementioned failure of most commercial programs to deal with system with feedback loops);

- it may shed lead new light on established topics, such as co-product allocation;

- it enables a further exploration of advanced topics, such as uncertainty analysis.

In conclusion, the aim of this book is to provide a comprehensive description of the present state of scientific knowledge of the computational structure of LCA.

1.2 Elements of LCA

The general ISO 14040 standard (ISO, 1997, p.2) defines LCA as the "compilation and evaluation of the inputs, outputs and the environmental impacts of a product system throughout its life cycle." The LCA technique is structured along a framework with a number of steps or activities in each of these steps. There are four phases:

- goal and scope definition;

- inventory analysis;

- impact assessment;

- interpretation.

A short summary of these phases follows.

Goal and scope definition deals with the clear and unambiguous formulation of the research question and the intended application of the answer that the LCA study is supposed to provide. Important elements of the goal and scope definition are the choice of the functional unit, the selection of product alternatives to be analysed, and the definition of the reference flows for each of the alternative systems.

The inventory analysis is concerned with the construction of these product systems. These systems are composed of unit processes, like industrial production, household consumption, waste treatment, transportation and so on. System boundaries and flow charts of linked unit processes are drawn for each alternative product system, and quantitative data as well as qualitative data for representativeness, etc. are collected during this phase. For those unit processes that are multifunctional, *i.e.* that provide more than one function, an allocation step is made. A final step of the inventory analysis is the aggregation of the emissions of chemicals and the extractions of natural resources over the entire product system, in such a way that a quantitative match with the system's reference flow is achieved. The final table of these aggregated emissions and extracted is referred to as the inventory table.

The result of the inventory analysis is often a long list with disparate entries, such as carbon dioxide, nitrogen oxides, chloromethane and mercury. The impact assessment aims to convert and aggregate these into environmentally relevant items. In particular, we mention here the step of characterisation, in which the inventory results are transformed into a number of contributions to environmental impact categories, such as global warming, acidification, and ecotoxicity. We also mention the optional normalisation in which the characterisation results are related to a reference value, such as the annual global extent of these impacts. We finally mention the weighting, in which priority weights are assigned to the characterisation or normalisation results, and which may result into one final score for each alternative product system.

During the course of the LCA, many choices and assumptions must be made. Moreover, uncertainty may be introduced with every data item. The interpretation phase deals with the meaning and robustness of the information obtained and processed in the previous phases. The interpretation may include comparisons with previously published LCA studies on similar products, uncertainty and sensitivity analyses, data checks, external comments, and much more. It is also the place in which a final judgement and decision is outspoken.

In using the LCA technique for carrying out an LCA study, one may distinguish several types of activities.

- There are activities, related to the design of the system, the collection of data, the making of assumptions and choices, and so on. This, for instance, includes steps like the drawing of system boundaries, the collection of process data, the choice of allocation method, and the choice of an impact assessment method.

- There are computational activities, related to transforming or combining data items into a certain result. For instance, emission data are related to the functional unit, aggregated over all unit processes in the system, multiplied with appropriate characterisation factors, and so on.

- There are activities that relate to the procedural embedding of an LCA project. Depending on the topic of study and the intended application, different stakeholders may be involved in certain ways. For certain applications, critical review by an independent expert is essential.

- There are activities, related to the planning of the LCA. For instance, one can start with a small-size LCA, to explore the potentials and bottlenecks, and then to reiterate the steps in a more complete way. Uncertainty analyses can give rise to further reiterations.

- There are activities, related to the reporting of an LCA. All types of requirements on what to report and how to report can be imposed to obtain transparent and reproducible reports.

The ISO-standards for LCA do not clearly separate these different types of activities. However, emphasis is, apart from the presentation of framework and the definition of terms, mainly on procedural embedding and reporting. Most importantly for this book, the ISO-standards for LCA do not cover the computational structure. One can easily confirm this by observing the absence of mathematical equations. This leaves a large degree of freedom for the present book. Many new technical terms will be introduced; examples are technology matrix and final demand vector. In fact, besides a presentation of the computational structure, this book aims to propose a standard nomenclature for a number of concepts; see Appendix B. Notation is also free, as there are no reserved symbols in the LCA-community (except perhaps one older proposal by Heijungs & Hofstetter (1995)). Throughout this book a consistent notation will be used. It is summarised in Appendix B as well. A number of new non-mathematical terms are introduced; we mention in particular hollow processes (Section 3.1), brands of economic flows (Section 3.4) and sleeping processes (Section 3.8). Finally, for a few terms that do occur in the ISO-standards, we have found reason to introduce a different meaning; here we mention reference flows (Section 3.7.2) and grouping (Section 8.1.6).

As already indicated, this book discusses the computational structure of LCA, without reference to the procedural embedding and without reference to the planning aspects. This means, for instance, that this book may well describe the mathematics of comparing product alternatives on the basis of a weighting procedure, while the ISO-standards state that such an activity is not appropriate. The point is that ISO's reluctance derives from procedural grounds, while the mathematics is in itself without problems. The mathematics remains valid even when someone decides to operate outside the ISO-framework, or when the ISO-standards are changed in this respect.

1.3 Background of the book

Most method-oriented texts on LCA focus on formulating guidelines (*cf.* Guinée *et al.* (2002)). In addition to that, there are many articles and reports in which specific topics are discussed, such as models for assessing impacts of acidification or data quality. There are only few texts in which the computational structure is discussed. To the extent that they are relevant for the present book, their material has been included. Important references in this respect include Projektgemeinschaft Lebensweg-bilanzen (1991), Heijungs *et al.* (1992), Möller (1992), Frischknecht *et al.* (1993), Heijungs (1994), Schmidt & Schorb (1995), Heijungs (1996), Heijungs (1997), Heijungs & Frischknecht (1998), Huele & van den Berg (1998) and Heijungs & Kleijn (2001).

In addition to that, other references that are relevant throughout the text are on linear algebra. Many texts, at various levels of sophistication and rigour, are available. Apostol (1969), Stewart (1973), Gentle (1997) and Harville (1997) provide good and accessible reviews. Albert (1972), Jennings & McKeown (1977) and Golub & Van Loan (1996) provide more specialised texts at an advanced level.

Finally, the topic of numerical analysis and computer algorithms is treated in many books, some emphasising the theoretical aspect and others providing easy-to-use computer codes. We have made use of the books by Jennings & McKeown (1977), Hamming (1986), Thisted (1988), Press *et al.* (1992) and Cheney & Kincaid (1999).

1.4   Structure of the book

1.4.1   Outline

This book discusses the computational structure of LCA. Much of the discussion will be directed to the computational aspects of inventory analysis. While many books on LCA would be structured along four core chapters, each of them dealing with one single phase of the LCA framework, this book presents the material in a different way. There is no chapter on goal and scope definition (although the reference flow is introduced in Section 2.1), and impact assessment and interpretation are treated in one single chapter (8).

Chapter 2 presents the basic computational model for inventory analysis. It introduces the representation of unit processes, economic flows and environmental flows, and it presents and solves the inventory problem: how to obtain the environmental flows associated with a functional unit.

Chapter 3 further develops the inventory analysis. We will see that the basic model falls short in many practical cases. This failure has to do with various complications that distort the ideal required for the basic model. These complications are, most importantly, cut-off and multifunctionality, the second type of complication giving rise to the allocation problem. This chapter also explores how the basic model works for a number of difficult situations.

Chapter 4 discusses advanced topics of the inventory analysis and is mainly intended for discussing very specific points. This chapter may be omitted without affecting the readability of the subsequent chapters. The same applies to Chapter 5 which ties the discussion to input-output analysis, a tool that is familiar in economics for more than sixty years and that shares certain features with LCA.

In Chapter 6, we abandon the idea of point estimates of data, and develop how the computational rules can be used to statistically deal with uncertainty. Both an analytical and a numerical treatment are included.

Chapter 7 discusses analytical explorations of the data on the basis of theoretical considerations. This leads to summary measures of the structure of the data and their dependencies.

All computational aspects beyond the inventory analysis are discussed in Chapter 8: impact assessment and interpretation.

Chapter 9 briefly explores a more general theory for LCA, in which the usual simplifying assumptions of linearity and steady state are abandoned.

A final chapter (10) is devoted to more information-technical topics: algorithms for the inversion of a matrix under special conditions, memory requirements, and so on.

Some sections contain special topics that can be omitted without distorting the readability of subsequent text. These sections are indicated with an asterisk (*).

The book assumes that the reader has a basic knowledge of matrix notation and manipulation. A concise review of matrix algebra is provided as an appendix (A). The first few chapters require a smaller background in mathematics than the chapters later on do. Especially Chapter 2, which discusses the basics, has been written in a more accessible way, to make sure that the basics can be understood by a wide audience. Chapter 3 is already more involved, and especially Chapter 6 requires quite some background.

1.4.2 Notation

In this book, a consistent notation will be employed throughout. Appendix B gives an overview of the most important symbols and the name of the concepts they represent. Furthermore, we have adhered to the convention that italic letters (like x) indicate scalars, that roman bold lowercase letters (like $\mathbf{x}$) indicate vectors and that roman bold uppercase letters (like $\mathbf{X}$) indicate matrices. A superscript T indicates the transpose of a vector or matrix, a superscript -1 the inverse of a matrix, a superscript $+$ the pseudoinverse of a matrix; see Appendix A for the definitions of these concepts. Other symbols that are placed after or on top of a symbol, like primes (x'), hats ($\hat{x}$), dots ($\dot{x}$) and tildes ($\tilde{x}$), are used to refer to another variable, and their meaning differs per occurrence. Sometimes, we will write a row vector for a column vector to save space, *e.g.* writing

$\mathbf{x} = \begin{pmatrix} 1 & 2 & 3 \end{pmatrix}^{\mathrm{T}}$ instead of $\mathbf{x} = \begin{pmatrix} 1 \\ 2 \\ 3 \end{pmatrix}$.

Chapter 2

The basic model for inventory analysis

In this chapter, the elementary formalism of the inventory analysis will be developed. It is based upon the simplifications that have been discussed by Guinée *et al.* (2002, p.III-15 *ff.*), *i.e.* a linear treatment of a steady-state situation. Approaches towards accounting for non-linearities and dynamic situations are discussed in Chapter 9. One could consider to start with the general model, and discuss the simplified model as a special case. This, however, would complicate the analytical treatment considerably, and it would moreover ignore that virtually all LCA studies, textbooks, software and databases are based on the simplified model. The general model is at present only an academic ideal, of which the practical applicability in concrete case studies is doubtful.

2.1 Representation of processes and flows

A first step in a formalised treatment is the construction of suitable system for the representation of quantified flows in connection with unit processes. For this, we introduce the notion of a linear space. A linear space is an abstract concept which allows us to uniquely represent a multidimensional data point as a simple vector with a definite value of each of the co-ordinates. See, *e.g.*, Apostol (1969) for an introduction into linear spaces.

For instance, consider a unit process (or process in short), say, production of electricity, which uses 2 litre of fuel to produce 10 kWh of electricity. Moreover, in doing so, it emits 1 kg of carbon dioxide and 0.1 kg of sulphur

11

dioxide. A linear space can now help us to describe this unit process in a very concise notation. We adopt the convention that the first dimension represents litre of fuel, that the second dimension represents kWh of electricity, that the third dimension represents kg of carbon dioxide and that the fourth dimension represents kg of sulphur dioxide. In term of linear spaces, the basis is

$$\begin{pmatrix} \text{litre of fuel} \\ \text{kWh of electricity} \\ \text{kg of carbon dioxide} \\ \text{kg of sulphur dioxide} \end{pmatrix} \tag{2.1}$$

Then the co-ordinates of the unit process production of electricity with respect to this basis is a simple vector

$$\mathbf{p} = \begin{pmatrix} -2 \\ 10 \\ 1 \\ 0.1 \end{pmatrix} \tag{2.2}$$

This will be referred to as the process vector for a particular unit process, in this case production of electricity.

Notice that we have written a minus sign in front of the 2 for the dimension that represents litre of fuel. The minus sign is a conventional indication for the direction of the flow. In Cartesian space, a negative x-coordinate indicates by convention a point at the left of the origin. Here, the negative co-ordinate indicates an input, while the other three positive coordinates indicate outputs. We emphasise the conventional nature of such a notation. In LCA, like in Cartesian geometry, a different choice leads to the same results when consistently followed.

Also notice that the vector that represents the unit process of electricity production has four co-ordinates in a definite order. We cannot interchange the elements of the vector, unless we change the order of the basis accordingly. Therefore, the order of the elements of the vector is fixed by convention as well. Again, this should be familiar from Cartesian geometry, where the first co-ordinate often represents the horizontal direction and the second the vertical direction.

A third type of convention is related to the choice of units. We might change the kg of kg of carbon dioxide into a mg. Of course, we can only do this if we change the co-ordinate 1 in the third row of the process vector into a $1,000,000$.

We will be involved with large systems comprising many different unit processes, like production of electricity, manufacturing of televisions, recycling of aluminium and transportation of tomatoes. A second step is therefore the representation of such a system of unit process. Let us consider a second unit process, say production of fuel. Suppose that for producing 100 litre of fuel, 50 litre of crude oil is needed, and that 10 kg of carbon dioxide and 2 kg of sulphur dioxide are emitted to the environment. A first thing to observe is that there is not yet an entry for crude oil in our four-dimensional linear space. A fifth dimension has therefore has to be added. Thus we change the basis into

$$
\begin{pmatrix}
\text{litre of fuel} \\
\text{kWh of electricity} \\
\text{kg of carbon dioxide} \\
\text{kg of sulphur dioxide} \\
\text{litre of crude oil}
\end{pmatrix}
\tag{2.3}
$$

and have to adapt the process vector for electricity production accordingly into

$$
\mathbf{p_1} =
\begin{pmatrix}
-2 \\
10 \\
1 \\
0.1 \\
0
\end{pmatrix}
\tag{2.4}
$$

The co-ordinates of the additional unit process, production of fuel, is then

$$
\mathbf{p_2} =
\begin{pmatrix}
100 \\
0 \\
10 \\
2 \\
-50
\end{pmatrix}
\tag{2.5}
$$

A particularly concise notation for representing the resulting system of unit process is

$$
\mathbf{P} = (\ \mathbf{p_1} \ | \ \mathbf{p_2} \ ) =
\begin{pmatrix}
-2 & 100 \\
10 & 0 \\
1 & 10 \\
0.1 & 2 \\
0 & -50
\end{pmatrix}
\tag{2.6}
$$

We will refer to this as the process matrix. Observe that a new convention is needed to express the fact that the first column represents the unit process of production of electricity, while the second column represents the unit process of production of fuel. Column vectors will be indicated as $\mathbf{p}_1$, $\mathbf{p}_2$ or $\mathbf{p}_j$ in general. An individual element of a process matrix can be referred to as $(\mathbf{P})_{ij}$ where i denotes the index of the row and j the index of the column. Observe that $(\mathbf{P})_{ij} = (\mathbf{p}_j)_i = p_{ij}$. In the example, i runs from 1 to 5 and j from 1 to 2. The process matrix is then said to be of dimension 5×2.

A third step is to partition the process matrix into two distinct parts: one representing the flows within the economic system, referred to as economic flows, and one representing the flows from and into the environment, referred to as environmental flows or environmental interventions or interventions for short. In the example, the first two rows, representing litre of fuel and kWh of electricity, are flows within the economic system, while the last three rows, representing kg of carbon dioxide, kg of sulphur dioxide and litre of crude oil are environmental flows. ISO (1997) speaks of product flows and elementary flows respectively, but the distinction between economic and environmental flows seems to be more popular. The partitioning leads to a partitioned matrix

$$\mathbf{P} = \left(\frac{\mathbf{A}}{\mathbf{B}} \right) = \left(\begin{array}{cc} -2 & 100 \\ 10 & 0 \\ \hline 1 & 10 \\ 0.1 & 2 \\ 0 & -50 \end{array} \right) \tag{2.7}$$

Although this partitioning is not needed *per se* for the representation of unit process or entire systems of unit processes, it is a convenient step. Furthermore, it will turn out to be needed in the following steps. The matrix $\mathbf{A}$ that represents the flows within the economic systems will be referred to as the technology matrix. Matrix $\mathbf{B}$ will be called the intervention matrix, because it represents the environmental interventions of unit processes. Partitioning in this way may lead to matrices and with an unequal number of rows. The number of columns of $\mathbf{A}$ and $\mathbf{B}$ is equal, and it is also equal to that of the unpartitioned process matrix $\mathbf{P}$.

A fourth step is more related to goal and scope definition than to inventory analysis. It involves the specification of the required performance of the system. In general, a reference flow ϕ will be determined as one way of fulfilling a functional unit that is quite arbitrarily chosen. For instance, a reference flow for this example could be 1000 kWh of electricity. The vector

$$\mathbf{f} = \begin{pmatrix} 0 \\ 1000 \end{pmatrix} \tag{2.8}$$

thus represents the set of economic flows that corresponds to this reference flow. Observe that we specify the complete set of economic flows, even though only one of these flows is the reference flow. The logic of using a co-ordinate system requires that we reserve an entry for every economic flow. In general, the only non-zero element of this vector, say the rth, is the reference flow:

$$f_i = \begin{cases} \phi & \text{if } i = r \\ 0 & \text{otherwise} \end{cases} \tag{2.9}$$

Vector $\mathbf{f}$ will be referred to as the final (or external) demand vector, because it is an exogenously defined set of economic flows of which we impose that the system produces exactly the given amount. Later on, in Section 3.4.2, we will discuss the case of comparing alternative products with more than one reference flow.

A final aspect of representation is the inventory table, *i.e.* the set of all environmental flows associated with the reference flow under consideration. How to find it will be the topic of the next section. For now, it suffices to discuss its notation. In the example co-ordinate system, we have three environmental flows. Even though some of these flows may be zero for a certain choice of $\mathbf{f}$, we need to reserve vector elements for each of these flows. We will proceed to define

$$\mathbf{g} = \begin{pmatrix} g_1 \\ g_2 \\ g_3 \end{pmatrix} \tag{2.10}$$

as a vector of environmental interventions, the inventory vector, where g_1 denotes the number of kg of carbon dioxide emitted by the total system, etc. The final demand vector and the inventory vector can be regarded as the aggregated external flows of the entire system. Stacking the two vectors

$$\mathbf{q} = \begin{pmatrix} \mathbf{f} \\ \mathbf{g} \end{pmatrix} = \begin{pmatrix} 0 \\ 1000 \\ g_1 \\ g_2 \\ g_3 \end{pmatrix} \tag{2.11}$$

provides an easy reference to this system vector.

2.2   The inventory problem and its solution

So far, we have only discussed the representation of unit processes, systems of unit processes, reference flows, and so on. We did not calculate anything yet. In particular, we did not yet discuss how to obtain the values of g_1, g_2 and g_3. A treatment of this leads to a discussion of what we will call the inventory problem.

The two unit processes produce 10 kWh of electricity and 100 litre of fuel respectively. The reference flow is 1000 kWh of electricity. Reference flow and flows produced by the unit process do not match. We see that unit processes 1 and 2 produce 10 and 0 kWh of electricity, while the final demand is 1000 kWh. Obviously, we need to scale up unit process 1 by a factor of 100 in order to satisfy the 1000 kWh required. But it is equally obvious that the fuel requirement by that process will be scaled up by the same factor of 100, into 200 litre of fuel. This leads to an upscaling of the second unit process by a factor of 2, so that it produces 200 litre of fuel. This then matches exactly with the required 200 litre of fuel by the first unit process. There is no surplus nor a shortage, hence the system's flow of fuel is 0, precisely as was required by the final demand vector.

Apart from the fact upscaling a unit process affects the economic flows, it affects the environmental flows in the same way. For instance, the emission of carbon dioxide by the first unit process is upscaled from 1 kg into 100 kg. For the second unit process it is upscaled from 10 kg into 20 kg. A total system-wide emission of carbon dioxide of 120 kg is therefore found. In other words, the hitherto unknown g_1 is found to be 120. For the other two elements of the inventory vector, similar calculations yield $g_2 = 14$ and $g_3 = -100$. Recall that the minus sign indicates an input, in this case extraction of 100 litre of crude oil.

A more formal treatment can now be given. First, we introduce a vector with scaling factors, the scaling vector, as a generalisation of the factors of 100 and 2. We will indicate this vector by $\mathbf{s}$ and write in the example case

$$\mathbf{s} = \begin{pmatrix} s_1 \\ s_2 \end{pmatrix} \tag{2.12}$$

For the first economic flow, fuel, a balance equation can be set up:

$$a_{11} \times s_1 + a_{12} \times s_2 = f_1 \tag{2.13}$$

In the concrete case, this amounts to

$$-2 \times s_1 + 100 \times s_2 = 0 \tag{2.14}$$

This equation cannot uniquely be solved for s_1 and s_2. But there is a second balance equation available, for the second economic flow, electricity:

$$a_{21} \times s_1 + a_{22} \times s_2 = f_2 \tag{2.15}$$

or with the coefficients inserted,

$$10 \times s_1 + 0 \times s_2 = 1000 \tag{2.16}$$

Simultaneous solution of these two equations yields

$$\mathbf{s} = \begin{pmatrix} s_1 \\ s_2 \end{pmatrix} = \begin{pmatrix} 100 \\ 2 \end{pmatrix} \tag{2.17}$$

A final step towards a generally applicable treatment is in terms of matrix solution. The system of equations

$$\begin{cases} a_{11} \times s_1 + a_{12} \times s_2 = f_1 \\ a_{21} \times s_1 + a_{22} \times s_2 = f_2 \end{cases} \tag{2.18}$$

can be written as

$$\begin{pmatrix} a_{11} & a_{12} \\ a_{21} & a_{22} \end{pmatrix} \begin{pmatrix} s_1 \\ s_2 \end{pmatrix} = \begin{pmatrix} f_1 \\ f_2 \end{pmatrix} \tag{2.19}$$

or even more concisely as

$$\mathbf{As} = \mathbf{f} \tag{2.20}$$

Given that the technology matrix $\mathbf{A}$ is known and that the final demand vector $\mathbf{f}$ is known, the balance equation can, under certain restrictions which are to be discussed in Section 2.4, be solved to yield the scaling vector $\mathbf{s}$:

$$\mathbf{s} = \mathbf{A}^{-1}\mathbf{f} \tag{2.21}$$

where $\mathbf{A}^{-1}$ denotes the inverse matrix of the technology matrix $\mathbf{A}$. In the example case, we have

$$\mathbf{A} = \begin{pmatrix} -2 & 100 \\ 10 & 0 \end{pmatrix} \tag{2.22}$$

and

$$\mathbf{A}^{-1} = \begin{pmatrix} 0 & 0.1 \\ 0.01 & 0.002 \end{pmatrix} \tag{2.23}$$

Straightforward multiplication yields

$$\mathbf{s} = \mathbf{A}^{-1}\mathbf{f} = \begin{pmatrix} 0 & 0.1 \\ 0.01 & 0.002 \end{pmatrix} \begin{pmatrix} 0 \\ 1000 \end{pmatrix} = \begin{pmatrix} 100 \\ 2 \end{pmatrix} \tag{2.24}$$

So, we have found a recipe to calculate the scaling vector for the unit processes in a system, such that the system-wide aggregation of economic flows exactly agrees with the final demand vector that represents the predetermined reference flow of the system. However, the inventory problem has not yet been solved completely, because the question was defined as to find the values of the system-wide aggregated environmental flows.

The scaling vector provides a direct clue to the final step in solving the inventory problem. We must recognise that scaling of a unit process affects both the economic flows and the environmental flows. For the first environmental flow, carbon dioxide, we have

$$g_1 = b_{11} \times s_1 + b_{12} \times s_2 \qquad (2.25)$$

In the concrete case, this amounts to

$$g_1 = 1 \times s_1 + 10 \times s_2 \qquad (2.26)$$

Inserting the values for s_1 and s_2, we find for g_1

$$g_1 = 1 \times 100 + 10 \times 2 = 120 \qquad (2.27)$$

More generally, we have

$$\left\{ \begin{array}{l} g_1 = b_{11} \times s_1 + b_{12} \times s_2 \\ g_2 = b_{21} \times s_1 + b_{22} \times s_2 \\ g_3 = b_{31} \times s_1 + b_{32} \times s_2 \end{array} \right. \qquad (2.28)$$

or in matrix notation

$$\mathbf{g} = \mathbf{Bs} \qquad (2.29)$$

In the example case, we have

$$\mathbf{B} = \begin{pmatrix} 1 & 10 \\ 0.1 & 2 \\ 0 & -50 \end{pmatrix} \qquad (2.30)$$

Matrix multiplication gives

$$\mathbf{g} = \mathbf{Bs} = \begin{pmatrix} 1 & 10 \\ 0.1 & 2 \\ 0 & -50 \end{pmatrix} \begin{pmatrix} 100 \\ 2 \end{pmatrix} = \begin{pmatrix} 120 \\ 14 \\ -100 \end{pmatrix} \qquad (2.31)$$

In principle, the inventory problem is now solved. There is a rule ($\mathbf{s} = \mathbf{A}^{-1}\mathbf{f}$) that yields the scaling vector given a technology matrix and a final demand vector. And there is a second rule ($\mathbf{g} = \mathbf{Bs}$) that yields the inventory vector given the intervention matrix and the scaling vector.

In certain situations, it may be useful to provide explicit formulations without matrix algebra. This leads to the following formulae:

$$\forall i : \sum_j a_{ij} s_j = f_i \tag{2.32}$$

for the balance equation, and

$$\forall k : g_k = \sum_j b_{kj} s_j \tag{2.33}$$

for the elements of the inventory vector, *i.e.* for the environmental interventions g_k.

An interesting substitution of variables can now be made. If the expression for the scaling factors is inserted in the expression for the environmental interventions, we find

$$\mathbf{g} = \mathbf{B}\mathbf{A}^{-1}\mathbf{f} \tag{2.34}$$

Matrix multiplication, like ordinary multiplication, is an associative operation, hence we may rewrite this as

$$\mathbf{g} = \left(\mathbf{B}\mathbf{A}^{-1}\right)\mathbf{f} \tag{2.35}$$

which we will write as

$$\mathbf{g} = \mathbf{\Lambda}\mathbf{f} \tag{2.36}$$

where we have defined the intensity matrix $\mathbf{\Lambda}$ as

$$\mathbf{\Lambda} = \mathbf{B}\mathbf{A}^{-1} \tag{2.37}$$

This notation makes clear that the matrix $\mathbf{\Lambda}$ can be evaluated for a particular system of unit processes, and then be applied to any final demand vector, thus to any reference flow that emanates from the system. In the example we have

$$\mathbf{\Lambda} = \begin{pmatrix} 0.1 & 0.12 \\ 0.02 & 0.014 \\ -0.5 & -0.1 \end{pmatrix} \tag{2.38}$$

This matrix can, for instance, be applied to

$$\mathbf{f} = \begin{pmatrix} 0 \\ 1000 \end{pmatrix}; \; \mathbf{f} = \begin{pmatrix} 0 \\ 1 \end{pmatrix}; \; \mathbf{f} = \begin{pmatrix} 10 \\ 0 \end{pmatrix};$$
$$\mathbf{f} = \begin{pmatrix} 10 \\ 1000 \end{pmatrix}; \; \mathbf{f} = \begin{pmatrix} -10 \\ 0 \end{pmatrix}; \; \text{etc.} \tag{2.39}$$

The meaning of these different types of final demand vectors will be discussed in Section 3.9. Using the matrix Λ implies that the scaling vector is not calculated. Even though the computation may be somewhat more efficient, knowledge of the intermediate results, in particular the scaling factors can provide a convenient tool for diagnosis of the results. Later on, in Section 2.6, we will also see that the scaling factors in some situations have a special meaning.

2.3   General formulation of the basic model for inventory analysis

The previous two sections have provided a view of the formalism and its rationale. But they have not provided a rigid formulation, and a scientific foundation is lacking anyway. This section provides such a general formulation. Readers interested in a more heuristical exposition of the computational structure of LCA may wish to defer the material in this section until they have gone through the other chapters, or they may decide to skip it all together.

The general formulation is based upon the principles of deductive logic: concepts are defined by formal definitions, *a priori* properties are assigned by axioms, and new properties are derived by lemmas or theorems, requiring a formal proof. Consequently, the following text is rather terse. Argumentations and illustrative examples are given in the previous sections.

We must first define the main objects of study, and postulate some of their properties. These include process vectors and matrices, the scaling vector, the final demand vector, as well as the property of linearity and additivity.

Definition 1 *A process vector* **p** *is a vector in a linear space of which the basis represents flows of goods, materials, services, wastes, substances, natural resources, land occupation, sound waves, and possibly other relevant items. The coefficients of this vector represent the amount of these items absorbed or produced by a particular unit process. A negative coefficient indicates an input of the process, a positive coefficient an output of the process, and a zero coefficient indicates that the item is not affected by the process. Two subsets of flows are distinguished: those which come from or go to another process (the economic flows), and those which come from or go to the environment (the environmental flows).*

Definition 2 *A process matrix* **P** *is a set of process vectors, juxtaposed to one another. It may be partitioned into a technology matrix* **A** *that represents the exchanges between processes, and an intervention matrix* **B** *that represents the exchanges with the environment.*

Axiom 1 *Any process vector* $\mathbf{p}_j$ *may be multiplied with an arbitrary constant* s_j. *In other words, processes represent linear technologies, and there are no effects of scale in production or consumption.*

Note that this axiom can in its turn be presented as a theorem when higher-level axioms are postulated; see Theorem 3 in Heijungs (1998).

Definition 3 *The constants* s_j *referred to in Axiom 1 may be stacked to form a scaling vector* **s**.

Axiom 2 *Flows may be aggregated over various processes, paying respect to the sign.*

Definition 4 *A final demand vector* **f** *is a vector of economic flows. The coefficients of this vector represent the amount of these items that a system under consideration should absorb or produce.*

With these basis ingredients, the inventory problem can be formulated according to Lemma 1.

Lemma 1 *Let* **A** *be the technology matrix of a given system. In order to let the system absorb or produce a final demand vector* **f**, *a scaling vector* **s** *should be found such that the condition*

$$\mathbf{As} = \mathbf{f} \tag{2.40}$$

is met.

Proof Applying a scaling vector **s** to the system produces or absorbs a vector of economic flows $\tilde{\mathbf{f}}$. For one arbitrary economic flow i, we have, from Axiom 1 and Axiom 2,

$$\tilde{f}_i = a_{i1} \times s_1 + a_{i2} \times s_2 + \cdots \tag{2.41}$$

As this applies for all economic flows, it follows that

$$\tilde{\mathbf{f}} = \mathbf{As} \tag{2.42}$$

The system thus produces or absorbs this amount. When it is imposed that the system produces or absorbs $\mathbf{f}$, one should find a scaling vector $\mathbf{s}$, such that

$$\tilde{\mathbf{f}} = \mathbf{f} \tag{2.43}$$

or equivalently

$$\mathbf{f} = \mathbf{As} \tag{2.44}$$

Q.E.D.

Theorem 1 *The condition* $\mathbf{As} = \mathbf{f}$ *referred to in Lemma 1, leads to a unique solution*

$$\mathbf{s} = \mathbf{A}^{-1}\mathbf{f} \tag{2.45}$$

provided that $\mathbf{A}$ *is square and non-singular.*

Proof Substituting the expression (2.45) for $\mathbf{s}$ into the condition (2.40) of Lemma 1, we have

$$\mathbf{As} = \mathbf{A}\left(\mathbf{A}^{-1}\mathbf{f}\right) = \left(\mathbf{AA}^{-1}\right)\mathbf{f} = \mathbf{f} \tag{2.46}$$

which shows that the expression for $\mathbf{s}$ indeed is a solution. The appearance of the -1 to indicate inversion is allowed only if $\mathbf{A}$ is square and non-singular. In that case, linear algebra teaches us that the solution is unique. Q.E.D.

Now, we proceed to define the inventory vector and the recipe how to find them.

Definition 5 *An inventory vector* $\mathbf{g}$ *is a vector of environmental flows. The coefficients of this vector represent the amount of these items that a system under consideration absorbs or produces.*

Theorem 2 *Let* $\mathbf{B}$ *be the intervention matrix of a given system. With a given scaling vector* $\mathbf{s}$, *the inventory vector* $\mathbf{g}$ *is given by*

$$\mathbf{g} = \mathbf{Bs} \tag{2.47}$$

Proof For one arbitrary environmental flow k, we have, from Axiom 1 and Axiom 2,

$$g_k = b_{k1} \times s_1 + b_{k2} \times s_2 + \cdots \tag{2.48}$$

As this applies for all environmental flows, Theorem 2 follows directly. Q.E.D.

This is, in fact, the entire axiomatic system for inventory analysis, at least for the basic case. Section 2.4 and Chapter 3 will discuss situations in which things are not so straightforward. In connection to Theorem 1, it may be noted that we have excluded the case that **A** is non-square or singular. In that case, there are two possibilities: either there is a solution, be it or not unique, that can be found by a different method; or there is not a solution, although there may be approximate solutions.

2.4   Some notes on the basic model

The basic model and its solution have been presented above for a very simple example case and in a generalised form using matrix notation. The main idea has been the systematic construction of a set of linear balance equations, one for each economic flow, with a number of scaling factors, one for each unit process. Matrix inversion has been introduced as a way to solve such a system of linear equations. However, it is not the only way to find a solution; see Section 4.1. Moreover, matrix inversion is a time and memory consuming operation, that is not easily accessible to those with insufficient mathematical training. It may under certain conditions be an operation that is numerically unstable, producing incorrect results; see Sections 6.6 and 10.2. Finally, in many situations, it is not directly applicable to LCA. Matrix inversion requires that the technology matrix is square and invertible. This is not automatically the case in situations involving

- cut-off of economic flows;

- multifunctional unit processes;

- a choice between alternative processes;

- closed-loop recycling.

How to adapt the matrix approach is described in Chapter 3. Furthermore, the approach outlined above (and in Chapters 3 and 4) start from the assumption of complete certainty, whereas it is for sure that process data are often uncertain to some degree. The treatment of uncertainties is discussed in Chapter 6. Finally, the assumption of linear scaling of processes as well as the effective neglect of temporal and spatial patterns are subjects for discussion in Chapter 9.

2.5   Geometric interpretation of inventory analysis*

Using the concepts of linear spaces suggests a link with Cartesian space, for which a geometric interpretation is readily available. Let us start with the economic flows. The basis for this subspace is

$$\begin{pmatrix} \text{litre of fuel} \\ \text{kWh of electricity} \end{pmatrix} \tag{2.49}$$

We can easily visualise the basis in a rectangular graph in which the first basis vector is $\hat{\mathbf{p}}_1 = \begin{pmatrix} 1 & 0 \end{pmatrix}^{\mathrm{T}}$, represents 1 litre of fuel, and is shown as a vector to the right. The second basis vector $\hat{\mathbf{p}}_1 = \begin{pmatrix} 0 & 1 \end{pmatrix}^{\mathrm{T}}$ represents 1 kWh of electricity and is shown as an upward vector.

Then, the unit process production of electricity, $\mathbf{p}_1$, can be represented as a vector, starting from the origin $\begin{pmatrix} 0 & 0 \end{pmatrix}^{\mathrm{T}}$ and ending in $\begin{pmatrix} -2 & 10 \end{pmatrix}^{\mathrm{T}}$. The unit process production of fuel, $\mathbf{p}_2$, also starts at the origin; it ends in $\begin{pmatrix} 100 & 0 \end{pmatrix}^{\mathrm{T}}$. The notion of a linear space implies that vectors can be added with the parallelogram rule, and that vectors may be multiplied with a scalar number. See Figure 2.1 for an illustration of the two unit processes $\mathbf{p}_1$ and $\mathbf{p}_2$ and their sumvector $\mathbf{p}_1 + \mathbf{p}_2$.

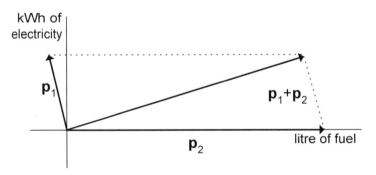

Figure 2.1: Geometric interpretation of two unit processes and their sum in a two-dimensional linear space representing two economic flows.

In the example case, we had two unit processes. There was one additional item: the final demand vector $\mathbf{f}$. It can be drawn in the graph as a vector starting from the origin and ending in $(\ 0 \quad 1000\ )^{\mathrm{T}}$. The inventory problem now consists of the question of finding the appropriate linear combination of unit process vectors, such that the resulting vector exactly coincides with the final demand vector. See Figure 2.2.

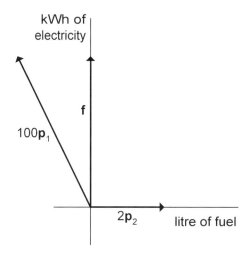

Figure 2.2: Geometric interpretation of how a linear combination of two unit processes $\mathbf{p}_1$ and $\mathbf{p}_2$ add up to the final demand vector $\mathbf{f}$.

A final step in the geometric interpretation is the addition of the environmental dimensions. However, the full example would require a graph in five dimensions. Therefore, we will restrict the illustration to only one environmental flow: kg of carbon dioxide. We extend the graph with a projected axis to represent depth. The basis of the new space is thus

$$\begin{pmatrix} \text{litre of fuel} \\ \underline{\text{kWh of electricity}} \\ \text{kg of carbon dioxide} \end{pmatrix} \qquad (2.50)$$

The basis vector is $\hat{\mathbf{p}}_3 = (\ 0 \quad 0\ |\ 1\ )^{\mathrm{T}}$ represents 1 kg of carbon dioxide. The first unit process, production of electricity, ends in the point with co-ordinates $(\ -2 \quad 10\ |\ 1\ )^{\mathrm{T}}$, the second one in $(\ 100 \quad 0\ |\ 10\ )^{\mathrm{T}}$. Now the system-wide aggregation. For the economic flows, this is the exogenously determined final demand vector. For the environmental flows, this is the inventory vector, which is not yet known. We could write these together as $\begin{pmatrix} 0 & 1000 & | & ? \end{pmatrix}^T$, where the question mark indicates that this co-ordinate is unknown. This three-dimensional aggregated vector has thus fixed values for the co-ordinates in the first and second dimension, but can temporarily assume any value for the co-ordinate in the third dimension. This corresponds with a straight line that passes through $\begin{pmatrix} 0 & 1000 & | & 1 \end{pmatrix}^T$, $\begin{pmatrix} 0 & 1000 & | & 0 \end{pmatrix}^T$, $\begin{pmatrix} 0 & 1000 & | & -1 \end{pmatrix}^T$, etc. See Figure 2.3. Now, we can interpret the inventory problem as finding a linear combination of the unit process vectors, such that the resulting vector falls on the line that is defined by the final demand vector.

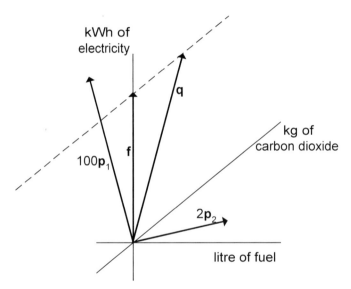

Figure 2.3: Geometric interpretation of the inventory problem as the problem of locating the point on an axis parallel to the axis that defines the environmental flow (kg of carbon dioxide) and passing through the point that is defined by the final demand vector **f**.

Recall that we have left out the environmental flows kg of sulphur dioxide and litre of crude oil for simplicity. If we add the fourth dimension, kg of sulphur dioxide, the final demand vector defines not a line in three dimensions, but a plane in four dimensions. And adding another dimension for litre of crude oil means that the final demand vector defines a three-dimensional object (a hyperplane) in five dimensions.

The final geometric interpretation of the inventory problem is thus to find a linear combination of the unit process vectors, such that the resulting vector falls on the hyperplane that is defined by the final demand vector, and to locate the exact co-ordinates of this resulting vector.

2.6 An interpretation of the scaling factors*

In Section 2.2 the notion of a scaling factor was introduced, as a factor which can serve to scale a unit process up or down. The idea is that a unit process is modeled as an activity that can be described with constant technical coefficients, *i.e.* representing a linear technology (*cf.* Axiom 1). The inventory problem was formulated as a geometrical problem in a linear space: find coefficients $s_1, s_2, \ldots$, such that a linear combination of the vectors that represent the economic part of the unit processes exactly matches with the final demand vector:

$$s_1\mathbf{p}_1 + s_2\mathbf{p}_2 + \cdots = \mathbf{f} \qquad (2.51)$$

This is only a valid interpretation if the vectors that represent the economic part of the unit processes and the final demand vector are indeed to be represented in the same linear space, and that the meaning of the co-ordinates, is therefore the same for unit processes and final demand. The scaling factors are then pure, *i.e.* dimensionless, numbers. There is, how-ever, an arbitrary element involved in the way a unit process is represented. If a unit process is given as $\mathbf{p}_j$, an arbitrary multiple $c \times \mathbf{p}_j$ could serve as an equally good representation. For instance, one can represent the unit process of electricity production per 10 kWh of electricity, per 100 kWh of electricity, per 1.23 kWh of electricity, and so on. And if this unit process in the original representation receives a scaling factor s_1, it would receive a scaling factor s_1/c in the revised representation. There is no preferred representation, and the scaling factors have no absolute meaning.

It is, however, possible to revise the scheme a bit (Heijungs, 1998). This starts with the acknowledgement that the descriptive data of a unit process are almost never recorded per 10 kWh, per 100 kWh, per 1.23 kWh, etc. A convenient way of measuring and straightforward recording of such data is per unit of time. The inputs and outputs of a unit process can be measured during an hour, a day, or a year. The descriptive data of a unit process then assume a form like 3.15×10^{10} kWh/year or 10 kWh/second. Then the step of rescaling the data of a unit process to a convenient size, such as 10 kWh, is no longer needed. The final demand vector, however, remains unaffected by such a redefinition of the basis of the linear space that defines the co-ordinates of the unit processes. It may still be 1000 kWh of electricity. This then leads to a readjustment of the meaning of the scaling factors. In the previous representation, a scaling factor of 100 meant that the arbitrarily rescaled unit process needed to be multiplied with an equally meaningless factor of 100. In the revised representation, the scaling factor bears a dimension: it is 100 second. Moreover, it can be interpreted as that the reference flow defined in the final demand vector requires that the unit process of electricity production is involved for 100 seconds. The electricity generator is thus allocated, so to speak, for 100 seconds to the function investigated.

The advantage of the new representation is three-fold:

- the step of scaling all unit process data to some convenient round number of output is not needed;

- the scaling factors receive a clear meaning;

- the process data can be entered in their actual extent, which may be convenient for the construction of databases that serve more purposes than LCA alone.

A disadvantage is that the basis that defines the co-ordinate system is not any longer universally applicable to both unit processes and final demand vector. This then also implies that the geometrical interpretation of Section 2.5 does not apply for this setup.

2.7   An interpretation of the intensity matrix*

In Section 2.2 we have introduced the matrix $\mathbf{\Lambda}$ as

$$\mathbf{\Lambda} = \mathbf{B}\mathbf{A}^{-1} \tag{2.52}$$

and given it the name intensity matrix. Some comments are in order: what does the matrix mean, and whence the name intensity matrix?

Let us first study the case of a technology matrix that consists of one number only. Let

$$\mathbf{A} = \begin{pmatrix} 10 \end{pmatrix} \tag{2.53}$$

denote that the system contains one process which produces 10 kWh of electricity. Let us assume that the intervention matrix is

$$\mathbf{B} = \begin{pmatrix} 1 \\ 0.1 \\ 0 \end{pmatrix} \tag{2.54}$$

which keeps to mean emission of 1 kg of carbon dioxide and of 0.1 kg of sulphur dioxide, and no in- or outflow of crude oil. Then

$$\mathbf{\Lambda} = \begin{pmatrix} 1 \\ 0.1 \\ 0 \end{pmatrix} \left(\; 10 \; \right)^{-1} = \begin{pmatrix} 0.1 \\ 0.01 \\ 0 \end{pmatrix} \tag{2.55}$$

Applying this to a (degenerate) final demand vector

$$\mathbf{f} = \left(\; 1 \; \right) \tag{2.56}$$

representing 1 kWh of electricity, we find

$$\mathbf{g} = \begin{pmatrix} 0.1 \\ 0.01 \\ 0 \end{pmatrix} \tag{2.57}$$

We thus see that the meaning of the coefficient 0.1 for Λ_{11} is that it represents the carbon dioxide intensity of electricity (in kg per kWh). Similarly, the coefficient 0.01 for Λ_{21} represents the sulphur dioxide intensity of electricity.

This suggest to assign to $\mathbf{\Lambda}$ the interpretation of a matrix of environmental intensity coefficients per unit of economic flow. However, this is too rapid a conclusion. Let us expand the technology matrix to the case of two processes and two flows:

$$\mathbf{A} = \begin{pmatrix} -2 & 100 \\ 10 & 0 \end{pmatrix} \tag{2.58}$$

Retrieving the intervention matrix

$$\mathbf{B} = \begin{pmatrix} 1 & 10 \\ 0.1 & 2 \\ 0 & -50 \end{pmatrix} \tag{2.59}$$

one finds

$$\mathbf{\Lambda} = \begin{pmatrix} 1 & 10 \\ 0.1 & 2 \\ 0 & -50 \end{pmatrix} \begin{pmatrix} -2 & 100 \\ 10 & 0 \end{pmatrix}^{-1} = \begin{pmatrix} 0.1 & 0.12 \\ 0.02 & 0.014 \\ -0.5 & -0.1 \end{pmatrix} \tag{2.60}$$

When applied to a final demand vector

$$\mathbf{f} = \begin{pmatrix} 0 \\ 1 \end{pmatrix} \tag{2.61}$$

again representing 1 kWh of electricity, we find

$$\mathbf{g} = \begin{pmatrix} 0.12 \\ 0.014 \\ -0.1 \end{pmatrix} \tag{2.62}$$

This is different from the **g** obtained for the case of one process and one flow, in spite of the fact that the specification of the process of electricity production is almost similar. The difference can be interpreted as an increase of environmental interventions: 0.02 kg extra emission of carbon dioxide, 0.004 kg extra emission of sulphur dioxide, and 0.1 litre extra extraction of crude oil (mind the removal of the minus sign for crude oil). The only difference in process specification is that the process of electricity production is assumed to consume fuel: 0.2 litre per 1 kWh electricity. We may also observe that the final demand vector that represents 1 litre of fuel,

$$\mathbf{f} = \begin{pmatrix} 1 \\ 0 \end{pmatrix} \tag{2.63}$$

yields

$$\mathbf{g} = \begin{pmatrix} 0.1 \\ 0.02 \\ -0.5 \end{pmatrix} \tag{2.64}$$

as inventory vector. Rescaled with a factor of 0.2, we find 0.02 kg of carbon dioxide, 0.004 kg of sulphur dioxide, and -0.1 litre of crude oil, which exactly accounts for the extra interventions of the delivery 1 kWh electricity. Thus, we arrive at an interpretation that the vector

$$\mathbf{\Lambda}_2 = \begin{pmatrix} 0.12 \\ 0.014 \\ -0.1 \end{pmatrix} \tag{2.65}$$

represents the system-wide (or cradle-to-grave) interventions of supplying 1 kWh of electricity, *i.e.* one unit of economic flow number 2. Similarly,

$$\mathbf{\Lambda}_1 = \begin{pmatrix} 0.1 \\ 0.002 \\ -0.5 \end{pmatrix} \tag{2.66}$$

represents the system-wide (or cradle-to-grave) interventions of supplying 1 litre of fuel, *i.e.* one unit of economic flow number 1 .

Hence, one may interpret a column of the intensity matrix as the system-wide interventions for supplying one unit of the good or service that is referred to by that column. Altogether, it seems reasonable to assign to $\mathbf{\Lambda}$ the interpretation of a matrix of system-wide environmental intensity coefficients per unit of economic flow. We refer to Section 3.8.2 for a longer discussion.

Chapter 3

The refined model for inventory analysis

The basic model for inventory analysis presented in Chapter 2 is appealing for reasons of simplicity and generality. In practice, however, things are more complicated for a variety of reasons, some of which have been listed already in Section 2.4. This calls for adaptations of the matrix approach. This chapter discusses the most important situations that lead to a readjustment of the model or its solution. It also shows how the formalism can be used for less trivial processes and flows, such as those relating to transport.

3.1 Cut-off

Cut-off refers to the case of incomplete systems, or formulated differently, to the case of systems with incomplete knowledge on product flows between processes. One may, for instance, know that production of electricity not only requires fuel, but that capital equipment, like a generator, is involved as well. But it may happen that one cannot find data on the production of new generators for replacement, or for the disposal of used generators.

Problems in data acquisition appear in almost every LCA. Most guidebooks provide criteria to decide when cut-off is permissible: for which processes, depending on the decision-context and on the product under investigation. This book will not provide or discuss such guidelines. The emphasis is on how cut-off is implemented in the matrix formalism of Chapter 2. As far as we are aware, this aspect of cut-off has been discussed only by Heijungs (1997, p.93–95). Another approach is to avoid the need for cut-off by estimating missing process data, for instance with input-output analysis; see Section 5.4.

3.1.1 Formulation of the problem

As an illustration, let us expand the basis of the economic part of the linear space by two new dimensions:

$$\begin{pmatrix} \text{litre of fuel} \\ \text{kWh of electricity} \\ \text{new generator} \\ \text{used generator} \end{pmatrix} \tag{3.1}$$

Further, let us assume that production of 10 kWh electricity uses 10^{-12} new generator and disposes 10^{-12} used generator. The new technology matrix is therefore

$$\mathbf{A} = \begin{pmatrix} -2 & 100 \\ 10 & 0 \\ -10^{-12} & 0 \\ 10^{-12} & 0 \end{pmatrix} \tag{3.2}$$

and the new final demand vector is

$$\mathbf{f} = \begin{pmatrix} 0 \\ 1000 \\ 0 \\ 0 \end{pmatrix} \tag{3.3}$$

Cut-off now implies that we have not included unit processes for the production of new generators and for the waste treatment of used generators. The system of unit processes should in principle be extended with unit processes to account for the production and waste treatment of generators. For certain reasons (*e.g.*, lack of appropriate data, lack of time for further data collection) it has been decided to restrict the technology matrix to the 4×2-dimensional one shown above.

It is immediately clear that the equation $\mathbf{As} = \mathbf{f}$ cannot be solved in this case. We can never find a scaling factor for unit process 1 that satisfies the final demand of 1000 kWh of electricity and at the same time has a zero for the third and fourth row. Formulated differently, we cannot invert a matrix that is rectangular, *i.e.* that is not square. We obviously could not have expected to be able to solve the set of balance equations in the case of missing unit processes. Apparently, cut-off implies more than a mere leaving out of unit processes from the system. A computational trick is needed in addition.

Following Heijungs (1997), we will discuss three such tricks:

- adding hollow processes to the technology matrix;

- replace non-zero numbers for flows that are cut-off by zeros;

- removing cut-off flows from the technology matrix.

The approaches mentioned discuss how to solve the computational problems of cut-off, not the problem itself, namely truncation due to an incomplete systems. Suh & Huppes (2000) suggest to link missing flows, that would be subject to cut-off, to the most similar commodity category in an input-output table to estimate the truncated portion of the interventions. This approach will be further discussed in Section 5.4.

3.1.2 Adding hollow processes to the technology matrix*

The problem of cut-off is essentially one of ignorance: process data are unknown for certain processes. Accepting the concept of cut-off implies that we accept that no information is added to the system. Adding no information could then be regarded equivalent as adding zeros. For instance, one could add a process to account for the production of new generators, with zeros for oil consumption, electricity use, CO_2 emission, etc. We will refer to this as a hollow process, because it is a process specification that contains no information at all. The only information it contains is that it produces a new generator:

$$
\mathbf{p_3} = \begin{pmatrix} 0 \\ 0 \\ 1 \\ \hline 0 \\ 0 \\ 0 \\ 0 \end{pmatrix} \tag{3.4}
$$

A similar hollow process can be added for the waste treatment of used generators. Here it is logical – though not essential – to add a minus sign, because the used generator is absorbed.

The two hollow processes can be added to the technology matrix:

$$\mathbf{A} = \begin{pmatrix} -2 & 100 & 0 & 0 \\ 10 & 0 & 0 & 0 \\ -10^{-12} & 0 & 1 & 0 \\ 10^{-12} & 0 & 0 & -1 \end{pmatrix} \tag{3.5}$$

The resulting system is square and invertible, and yields a scaling vector

$$\mathbf{s} = \begin{pmatrix} 100 \\ 2 \\ 0 \\ 0 \end{pmatrix} \tag{3.6}$$

We may note that this approach leads to larger technology matrices. This will in most cases not present serious computer problems.

3.1.3 Replacing non-zero numbers for flows that are cut-off by zeros*

A second approach is to replace the non-zero numbers for flows that are cut-off by zeros. In itself, this does not solve the problem of rectangularity, so a further step of removing the cut-off flows from the system is needed if one wants to obtain a square technology matrix. The procedure for removing cut-off flows from the technology matrix is treated hereafter. Obviously, it is not necessary to replace non-zero numbers by zeros when the numbers are removed anyhow. Therefore, we will restrict this discussion to showing how to remove flows, without the need to discuss how to insert zeros.

3.1.4 Removing cut-off flows from the technology matrix

The third solution is to single out the economic flows that are to be cut-off by introducing a further partitioning of the technology matrix and the final demand vector. That is, we write

$$\mathbf{A} = \left(\frac{\mathbf{A'}}{\mathbf{A''}} \right) = \begin{pmatrix} -2 & 100 \\ 10 & 0 \\ \hline -10^{-12} & 0 \\ 10^{-12} & 0 \end{pmatrix} \tag{3.7}$$

for the technology matrix and

$$\mathbf{f} = \left(\frac{\mathbf{f}'}{\mathbf{0}} \right) = \begin{pmatrix} 0 \\ 1000 \\ 0 \\ 0 \end{pmatrix} \tag{3.8}$$

and for the final demand vector. Then we solve the inventory part for the upper part of the system of equations:

$$\mathbf{A}'\mathbf{s} = \mathbf{f}' \tag{3.9}$$

which of course leads to the solution previously discussed:

$$\mathbf{s} = \mathbf{A}'^{-1}\mathbf{f}' = \begin{pmatrix} 100 \\ 2 \end{pmatrix} \tag{3.10}$$

and are able to find once more

$$\mathbf{g} = \mathbf{B}\mathbf{s} = \begin{pmatrix} 120 \\ 14 \\ -100 \end{pmatrix} \tag{3.11}$$

In addition, we can calculate the size of the neglected economic flows due to cut-off:

$$\mathbf{f}'' = \mathbf{A}''\mathbf{s} = \begin{pmatrix} -10^{-10} \\ 10^{-10} \end{pmatrix} \tag{3.12}$$

In other words, cut-off implies giving up the strict requirement of fulfilling the final demand vector $\mathbf{f}$. The modeled system produces an array of economic flows that does not fully agree with the final demand. It may be referred to as the final supply vector $\tilde{\mathbf{f}}$:

$$\tilde{\mathbf{f}} = \left(\frac{\mathbf{f}'}{\mathbf{f}''} \right) = \left(\frac{\mathbf{A}'}{\mathbf{A}''} \right)\mathbf{s} \tag{3.13}$$

or simply

$$\tilde{\mathbf{f}} = \mathbf{A}\mathbf{s} \tag{3.14}$$

In the example, we have

$$\tilde{\mathbf{f}} = \begin{pmatrix} 0 \\ 1000 \\ -10^{-10} \\ 10^{-10} \end{pmatrix} \tag{3.15}$$

There is a discrepancy between imposed final demand and obtained final supply, and we may write

$$\mathbf{d} = \tilde{\mathbf{f}} - \mathbf{f} \tag{3.16}$$

for this discrepancy vector. In the example, we have

$$\mathbf{d} = \begin{pmatrix} 0 \\ 0 \\ \hline -10^{-10} \\ 10^{-10} \end{pmatrix} \tag{3.17}$$

We will see in subsequent sections that there are more situations in which a non-zero discrepancy between final demand and final supply occurs.

A final remark in relation to partitioning the technology matrix and final demand vector relates to the order of the economic flows. The partitioning shown above requires that the flows that will be cut off are grouped together on the lower end of the list of economic flows. In general, this will not be the case. However, a simple change of the order of the flows can always be made, provided that it is carried out consistently in all vectors and matrices.

3.1.5   Brief discussion

Three methods for dealing with flows that have been decided to be cut-off have been described. We have a strong preference for the third approach – removing flows that have been cut-off from the technology matrix – as it is the only one that does not encourage one to generate wrong information, and the only one that allows for the calculation of a discrepancy vector. Inserting hollow processes and removing non-zero numbers by zeros come down to purposely corrupting the information content. Removing cut-off flows from the technology matrix is not the same as throwing away information: it only means that the technology matrix $\mathbf{A}$ is split into a "solvable" part $\mathbf{A}'$ and an "unsolvable" part $\mathbf{A}''$, and that the information on cut-off flows is shifted to the unsolvable part. It is precisely this aspect which enables the possibility of computing the discrepancy vector.

Even though our preference for handling cut-off is for the third approach, an even better idea is to avoid cut-off by estimating missing flows with other tools, most notably input-output analysis; see Section 5.4.

3.2   Multifunctionality and allocation

In industrial practice, it is frequently the case that a unit process produces more than one valuable product or material. For instance, refineries produce petrol, kerosene, heavy oil, diesel, and so on. More examples are the simultaneous production of chlorine and caustic soda, or the production of milk, meat, hides, calves and sperm by cow breeding. Unit processes which serve more than one function will be referred to as multifunctional unit processes.

We may categorise multifunctional unit processes along at least two lines. The first categorisation is according to the direction of the functions:

- unit processes that produce two or more valuable outputs (co-production, *e.g.*, production of chlorine and caustic soda);

- unit processes that treat two or more waste inputs (combined waste treatment, *e.g.*, incineration of plastic bottles and paper bags);

- unit processes that treat one waste input and that produce one valuable output (recycling, *e.g.*, upgrading of used newspapers into corrugated board);

- unit processes that serve three or more valuable functions (most general case, *e.g.*, as in a refinery).

A second categorisation relates to the inevitability of the multifunctionality:

- unit processes of which the functions are causally coupled (joint production, *e.g.*, production of chlorine and caustic soda);

- unit processes of which the functions are deliberately coupled (combined production, *e.g.*, transportation of passengers and cargo by an aeroplane).

Regardless of the categorisation, we can start to investigate the representation of such multifunctional process and the problems they create in finding a solution of the balance equations. After all, regarding the input of waste as delivery of a valuable service, waste treatment (see also Section 3.7.6), stresses that combined waste treatment can in fact be regarded as co-production of several waste treatment services. Likewise, recycling can be regarded as co-producing a waste treatment service and a secondary material. Furthermore, the mathematical treatment of unit processes as linear

technologies with fixed coefficients turns the distinction between joint production and combined production into a metaphysical one, with no consequences on the level of the analysis.

3.2.1   Formulation of the problem

Assume that the production of electricity takes place in a modern plant
which co-generates heat, for instance for use in municipal or industrial
heating systems. We add an extra economic flow to the basis to account
for MJ of heat:

$$\begin{pmatrix} \text{litre of fuel} \\ \text{kWh of electricity} \\ \text{MJ of heat} \\ \hline \text{kg of carbon dioxide} \\ \text{kg of sulphur dioxide} \\ \text{litre of crude oil} \end{pmatrix} \tag{3.18}$$

If we assume that 10 kWh of electricity is co-produced along with 18 MJ
of heat, the new process vector for production of electricity becomes

$$\mathbf{p_1} = \begin{pmatrix} -2 \\ 10 \\ 18 \\ \hline 1 \\ 0.1 \\ 0 \end{pmatrix} \tag{3.19}$$

Assuming no changes in the rest of the system, and maintaining the same
reference flow, 1000 kWh of electricity, we find

$$\mathbf{A} = \begin{pmatrix} -2 & 100 \\ 10 & 0 \\ 18 & 0 \end{pmatrix} \text{ and } \mathbf{f} = \begin{pmatrix} 0 \\ 1000 \\ 0 \end{pmatrix} \tag{3.20}$$

One could propose to write the solution to the inventory problem in this
case as normal:

$$\mathbf{s} = \mathbf{A}^{-1}\mathbf{f} \tag{3.21}$$

However, the technology matrix $\mathbf{A}$ is in this case not square: it has 3
rows and 2 columns, and is therefore rectangular. Such a matrix cannot
be inverted, hence the previously derived recipe for solving the inventory
problem cannot be used.

To clarify the situation, we will write down the set of balance equations. They are

$$\begin{cases} -2 \times s_1 + 100 \times s_2 = 0 \\ 10 \times s_1 + 0 \times s_2 = 1000 \\ 18 \times s_1 + 0 \times s_2 = 0 \end{cases} \qquad (3.22)$$

There are three equations in two unknowns. Such a system of equations is referred to as an overdetermined system of equations. An overdetermined system of equations can in general not be solved. In the example, the second equation leads to $s_1 = 100$, while the third equation gives $s_1 = 0$. This clearly demonstrates that an overdetermined system of equations easily contains mutually contradictory equations, and that a solution to such a system cannot be found. Observe that an overdetermined system may, but need not, contain mutually contradictory equations. In Section 3.5 we will discuss a case in which the equations are consistent with one another.

The occurrence of multifunctional unit processes in a system of processes leads to problems in solving the balance equations. Observe that there is no problem whatsoever in representing a multifunctional process, nor in constructing a system with multifunctional processes. The problem only appears when one tries to solve the equations. We nevertheless will refer to this situation as the multifunctionality problem. Another popular term for it is the allocation problem.

We will approach the multifunctionality problem from an analytical point of view. This means that the starting point is the overdeterminedness of the system of equations, which is reflected in the rectangularity of the technology matrix. Recall that the previous section on cut-off contained another rectangular matrix. There, rows were removed through a cut-off procedure, so that a square invertible matrix resulted. One way to deal with the multifunctionality problem is therefore once more the removal of rows from the technology matrix. But there is another approach: one could also increase the number of columns of the technology matrix. We will discuss both options, and within the second option present two entirely different ways of adding columns. We will start with this option of expanding the number of columns.

3.2.2 The substitution method

In the case of combined heat and power generation, the demand of 1000 kWh of electricity, leads to production of 18 MJ of heat as well. It may now be argued that this extra 18 MJ of heat, which is available anyhow, will cause that the stand-alone generation of heat, for domestic or industrial purposes, will decline with exactly 18 MJ. Thus the co-production of 18 MJ of heat may be argued to substitute the stand-alone production of 18 MJ of heat. The availability of the co-product avoids, so to say, the stand-alone production. This latter process can then be seen as an avoided process. The method in which the multifunctionality problem is solved with this principle will be referred to as the substitution method, although variations, such as avoided burdens method or subtraction method are encountered as well. We will first discuss and illustrate the computational details of this method.

The substitution method requires that a stand-alone unit process be specified to account for the avoided economic flow. In the example, the avoided flow is MJ of heat. Let us suppose that the stand-alone process is as follows: 90 MJ of heat is produced using 5 litre of fuel and emitting 3 kg of carbon dioxide. The matrix representation of the new technology matrix is therefore

$$\mathbf{A'} = \begin{pmatrix} -2 & 100 & -5 \\ 10 & 0 & 0 \\ 18 & 0 & 90 \end{pmatrix} \tag{3.23}$$

This matrix can be inverted. Multiplication of the inverse with the final demand vector yields the scaling vector:

$$\mathbf{s'} = \begin{pmatrix} 100 \\ 1 \\ -20 \end{pmatrix} \tag{3.24}$$

Later on, when we have introduced the concept of brands, we will have to make some comments on the construction of the square matrix above (Section 3.4.4).

Comparison with the original scaling vector $\mathbf{s} = \begin{pmatrix} 100 & 2 \end{pmatrix}^T$, for the system without co-production, shows three remarkable features:

- The scaling factor for the unit process production of electricity is unaffected. In general, this will be the case: the multifunctional process itself is needed in the same proportion, as the reference flow still requires the same amount of one of these flows.

- The scaling factor for production of fuel has decreased. Such a thing is also quite normal. The unit process stand-alone production of heat needs fuel, so the avoidance of this unit process can be seen to avoid the need for fuel.

- The scaling factor for the unit process stand-alone production of heat is negative. This unit process does not participate in the system, but it is subtracted from the system, which is the same as including it in a negative way. However, it may be the case that stand-alone production of heat is actually somewhere part of the life cycle, and that the substitution method makes that it is involved to a lesser extent. In general, therefore, the scaling factor of the avoided process is smaller than that of the system without co-production.

The intervention matrix is also expanded to include an extra column for the avoided process of stand-alone heat production. It is as follows:

$$\mathbf{B'} = \begin{pmatrix} 1 & 10 & 3 \\ 0.1 & 2 & 0 \\ 0 & -50 & 0 \end{pmatrix} \tag{3.25}$$

Multiplication of the scaling vector with this new intervention matrix gives

$$\mathbf{g} = \mathbf{B's'} = \begin{pmatrix} 50 \\ 12 \\ -50 \end{pmatrix} \tag{3.26}$$

We see that, compared to the original system without co-production of heat, where $\mathbf{g} = (\begin{array}{ccc} 120 & 14 & -100 \end{array})^{\mathrm{T}}$, all environmental interventions have decreased (in absolute value). In general, some environmental interventions will remain unaffected. It may also happen that certain environmental interventions are avoided while they were not involved in the original system. This for instance happens when waste incineration is assumed to co-produce electricity, and when the avoided electricity is assumed to be of nuclear or partly nuclear origin. In that case, there will be negative releases of certain radioactive isotopes.

In contrast to the case of cut-off discussed above, we cannot calculate a final supply vector with

$$\tilde{\mathbf{f}} = \mathbf{As} \tag{3.27}$$

The reason is that we do not have a vector $\mathbf{s}$ that can be applied to the original system. Only the vector $\mathbf{s'}$ for the expanded system can be calculated. Obviously, lack of a final supply vector implies lack of a discrepancy vector as well.

So far the basics of the substitution method. There remains a number of minor points, however. One of the largest problem of the substitution method is the question what exactly is to be chosen as the avoided process.

Unit processes that are modeled as being part of the life cycle can to some extent be verified to be really part of the life cycle. One can trace back where a certain material came from, and one can follow where a certain product will be delivered. It is principally impossible, however, to trace back materials or to follow products which are not there for reasons of being avoided. This is an important problem in applying the substitution method, but it is not a problem that ought to be discussed in relation to its computational structure.

Another point has to do with the complication that there is not always a stand-alone avoided process. It may happen that the avoided process itself is a multifunctional process itself. In that case, another unit process should be added to the system to account for the co-product of the avoided process. Such a doubly-avoided unit process can simply be accommodated with the computational rules described above. In practice, the existence of multifunctional avoided processes may of course lead to an ever-increasing system with more and more unit processes which have little or nothing to do with the product life cycle under study. This is more a philosophical than a computational issue.

An important modification to the simple scheme presented above is related to the fact that it often happens that a certain economic flow substitutes another economic flow, even though the two are not exactly identical. Assume, for instance, that the heat obtained from the co-production process is of quality X, *e.g.*, of a certain energy content, or available at a certain location, or available at a certain time, while the stand-alone production of heat produces heat of a quality Y, *e.g.*, with a slightly different energy content, or at a different location, or at a different time. We then need four rows to represent the technology matrix

$$\mathbf{A}' = \begin{pmatrix} -2 & 100 & -5 \\ 10 & 0 & 0 \\ 18 & 0 & 0 \\ 0 & 0 & 90 \end{pmatrix} \qquad (3.28)$$

in a basis of which the economic flows are

$$\begin{pmatrix} \text{litre of fuel} \\ \text{kWh of electricity} \\ \text{MJ of heat of quality } X \\ \text{MJ of heat of quality } Y \end{pmatrix} \qquad (3.29)$$

This matrix is again rectangular, hence it cannot be inverted. But we can make it square and invertible by explicitly addressing the equivalency or inequivalency of the two types of heat. If one assumes that the difference in quality is irrelevant, the two flows are assumed to be indistinguishable and hence the two rows can be merged, yielding

$$\mathbf{A'} = \begin{pmatrix} -2 & 100 & -5 \\ 10 & 0 & 0 \\ 18 & 0 & 90 \end{pmatrix} \tag{3.30}$$

with respect to a basis

$$\begin{pmatrix} \text{litre of fuel} \\ \text{kWh of electricity} \\ \text{MJ of heat of quality } X \text{ or } Y \end{pmatrix} \tag{3.31}$$

If one, on the other hand decides that the quality difference is relevant, a correction factor may be applied. Suppose, for instance, that we decide that heat of quality X is considered to be 10% more valuable than heat of quality Y, for instance on the basis of a higher energy content or availability at a more profitable place or time. This then means that 1 MJ of heat of quality X is assumed to be equivalent to 1.1 MJ of heat of quality Y. We can then merge the two flows and hence rows, applying a correction factor to account for the 10% difference. This yields

$$\mathbf{A'} = \begin{pmatrix} -2 & 100 & -5 \\ 10 & 0 & 0 \\ 18 & 0 & 81.8 \end{pmatrix} \tag{3.32}$$

with respect to a basis

$$\begin{pmatrix} \text{litre of fuel} \\ \text{kWh of electricity} \\ \text{MJ of heat of quality } X \end{pmatrix} \tag{3.33}$$

Obviously, such a factor is always needed when the two flows that are assumed to substitute one another are expressed in different units. For instance, when there is no difference in quality, 1 kWh of electricity is equivalent to 3.6 MJ of electricity. The factor thus accounts for differences in quality and differences in unit. We will refer to it as the exchange factor. Using the substitution method with an exchange factor that differs from 1 is sometimes referred to as applying value-corrected substitution.

3.2.3   The partitioning method

The second approach towards obtaining an invertible technology matrix
starts from the recognition that it is the multifunctionality of a unit process which causes the problem. Hence, splitting a multifunctional process
into a number of independent monofunctional processes should provide a
cure. This splitting will be referred to as partitioning, although the term
allocating is frequently seen, and although there is not a connection with
the partitioning of a matrix as we have run across earlier. The disadvantage
of the term allocation is, however, that the entire multifunctionality problem is often referred to as the allocation problem as well. In this book, we
try to avoid the confusion that may arise from such double use of the word
allocation. Moreover, economists use the term allocation in connection to
a host of other phenomena, such as budget allocation.

In the partitioning method, a multifunctional unit process is divided
into a number of unit processes, such that each of the resulting unit processes is monofunctional. In the case of combined production of electricity
and heat, two unit processes will be formed: one producing kWh of electricity and one producing MJ of heat. That is, the process vector

$$\mathbf{p}_1 = \begin{pmatrix} -2 \\ 10 \\ 18 \\ \hline 1 \\ 0.1 \\ 0 \end{pmatrix} \tag{3.34}$$

is split into two process vectors

$$\mathbf{p}_{1a} = \begin{pmatrix} \lambda_{1a} \times -2 \\ 10 \\ 0 \\ \hline \mu_{1a} \times 1 \\ \mu_{2a} \times 0.1 \\ 0 \end{pmatrix} \tag{3.35}$$

and

$$\mathbf{p}_{1b} = \begin{pmatrix} \lambda_{1b} \times -2 \\ 0 \\ 18 \\ \hline \mu_{1b} \times 1 \\ \mu_{2b} \times 0.1 \\ 0 \end{pmatrix} \tag{3.36}$$

where the coefficient λ_{1a}, λ_{1b}, μ_{1a}, μ_{1b}, μ_{2a} and μ_{2b} are allocation factors which achieve an allocation of the economic flow litre of fuel and the economic flows kg of carbon dioxide, kg of sulphur dioxide over the two newly created unit processes. At almost all places where the partitioning method is described or used, the allocation factors are chosen such that they lie between 0 and 1:

$$\left\{ \begin{array}{l} 0 \le \lambda_{1a} \le 1 \\ 0 \le \lambda_{1b} \le 1 \\ 0 \le \mu_{1a} \le 1 \\ 0 \le \mu_{1b} \le 1 \\ 0 \le \mu_{2a} \le 1 \\ 0 \le \mu_{2b} \le 1 \end{array} \right. \tag{3.37}$$

and that they add up to 1 per flow:

$$\left\{ \begin{array}{l} \lambda_{1a} + \lambda_{1b} = 1 \\ \mu_{1a} + \mu_{1b} = 1 \\ \mu_{2a} + \mu_{2b} = 1 \end{array} \right. \tag{3.38}$$

In that latter case, one has

$$\mathbf{p}_{1a} + \mathbf{p}_{1b} = \mathbf{p}_1 \tag{3.39}$$

and one can truly speak of a partitioning of the flows of the original multifunctional process among the new monofunctional processes. This is often referred to as the 100%-rule: the sum of the partitioned monofunctional processes is equal to the unpartitioned multifunctional process from which they are derived. A further simplification is that the allocation factors are almost always equal within one individual monofunctional unit process:

$$\left\{ \begin{array}{l} \lambda_{1a} = \mu_{1a} = \mu_{2a} \\ \lambda_{1b} = \mu_{1b} = \mu_{2b} \end{array} \right. \tag{3.40}$$

Under these conditions, the number of independently variable allocation factors reduces to the number of functions provided by the multifunctional unit process minus one. We loose one degree of freedom due to the constraint that the sum equals 1. In the example case, there were two functions, electricity and heat, and therefore one allocation factor λ_a suffices:

$$\mathbf{p}_{1a} = \left(\begin{array}{c} \lambda_a \times -2 \\ 10 \\ 0 \\ \lambda_a \times 1 \\ \lambda_a \times 0.1 \\ 0 \end{array} \right) \tag{3.41}$$

and

$$\mathbf{p}_{1b} = \begin{pmatrix} (1 - \lambda_a) \times -2 \\ 0 \\ \dfrac{18}{(1 - \lambda_a) \times 1} \\ (1 - \lambda_a) \times 0.1 \\ 0 \end{pmatrix} \tag{3.42}$$

As an example, let us construct the new technology matrix with the choice $\lambda_a = 0.7$.

$$\mathbf{A}' = \begin{pmatrix} 0.7 \times -2 & 100 & 0.3 \times -2 \\ 10 & 0 & 0 \\ 0 & 0 & 18 \end{pmatrix} \tag{3.43}$$

Matrix inversion is now possible and yields the scaling factors

$$\mathbf{s}' = \begin{pmatrix} 100 \\ 1.4 \\ 0 \end{pmatrix} \tag{3.44}$$

Comparison with the original scaling factors $\mathbf{s} = \begin{pmatrix} 100 & 2 \end{pmatrix}^\mathsf{T}$, for the system without co-production, shows the following:

- The scaling factor for the unit process production of electricity is unaffected. In general, this will be the case: the multifunctional process itself is needed in the same proportion, as the reference flow still requires the same amount of one of these flows.

- The scaling factor for production of fuel has decreased. Such a thing is also quite normal. The monofunctional unit process production of electricity needs less fuel, and this decreased need for fuel is reflected in a smaller scaling factor for production of fuel.

- The scaling factor for the monofunctional unit process production of heat is zero. This unit process does not participate in the system, not is a positive way, but also not in a negative way. It is a sleeping process (*cf.* Section 3.8.3) that could have been left out of the system altogether.

Let us now study what happens with the environmental flows. In the intervention matrix, the first column will be split as well. Using $\lambda_a = 0.7$

it becomes

$$\mathbf{B}' = \begin{pmatrix} 0.7 \times 1 & 10 & 0.3 \times 1 \\ 0.7 \times 0.1 & 2 & 0.3 \times 0.1 \\ 0.7 \times 0 & -50 & 0.3 \times 0 \end{pmatrix} \tag{3.45}$$

Multiplication with the new scaling vector gives the environmental interventions:

$$\mathbf{g} = \mathbf{B}'\mathbf{s}' = \begin{pmatrix} 84 \\ 9.8 \\ -70 \end{pmatrix} \tag{3.46}$$

Let us again make a comparison with the environmental interventions $\mathbf{g} = \begin{pmatrix} 120 & 14 & -100 \end{pmatrix}^{\mathrm{T}}$ in the case without co-production of heat. We see that all environmental flows are lower (in an absolute sense) than originally. More generally, certain interventions will be reduced while others will remain unchanged.

Notice that, as for the substitution method, we cannot calculate a separate expression for the final supply vector. Hence, there is no sensible expression for the discrepancy vector.

This concludes the basic description of the partitioning method. Recall that a large methodological problem with the substitution method was the specification of the avoided process. With the partitioning method, a different type of specification problem exists, namely the choice of the allocation factors λ and μ. Main strategies for this include parameters that relate to the chemical or physical causality, such as mass, energy-content or exergy, and parameters that relate to the economic or societal causality, such as costs or proceeds. Regardless the principles used for setting the allocation factors, the computational formulae are the same. Therefore, it is not necessary to devote separate sections to allocation on a physical basis and allocation on an economic basis.

3.2.4 The surplus method*

The third approach towards solving the multifunctionality problem is highly related to the case of cut-off discussed before. It starts with the assumption that some unit processes operate mainly for one of the functional flows. This is then often referred to as the main flow, while the others could be described as secondary flows or minor flows. Especially in a rapid LCA, one could decide to treat the multifunctional process as if it has been designed so as to produce this main flow, and that all minor flows are gratefully accepted, but that they will not receive any environmental burdens.

Effectively, the minor flows are removed from the balance equations. The technology matrix is partitioned into a part for which balance equations are constructed and a part for which this is not the case. In the example:

$$\mathbf{A} = \left(\frac{\mathbf{A'}}{\mathbf{A''}} \right) = \left(\begin{array}{cc} -2 & 100 \\ 10 & 0 \\ \hline 20 & 0 \end{array} \right) \tag{3.47}$$

and

$$\mathbf{f} = \left(\frac{\mathbf{f'}}{\mathbf{0}} \right) = \left(\begin{array}{c} 0 \\ 1000 \\ \hline 0 \end{array} \right) \tag{3.48}$$

and the balance equations are restricted to the upper part:

$$\mathbf{A's} = \mathbf{f'} \tag{3.49}$$

which obviously yield the old solution:

$$\mathbf{s} = \mathbf{A'}^{-1}\mathbf{f'} = \left(\begin{array}{c} 100 \\ 2 \end{array} \right) \tag{3.50}$$

This scaling vector can be applied to the intervention matrix, which is the same as before:

$$\mathbf{B} = \left(\begin{array}{cc} 1 & 10 \\ 0.1 & 2 \\ 0 & -50 \end{array} \right) \tag{3.51}$$

Obviously, the original environmental interventions are retrieved:

$$\mathbf{g} = \mathbf{Bs} = \left(\begin{array}{c} 120 \\ 14 \\ -100 \end{array} \right) \tag{3.52}$$

We could not have expected differently. The surplus method effectively comes down to ignoring the fact that useful flows are co-produced.

Interestingly, one may bring the surplus method in the framework of the partitioning method by splitting the multifunctional unit process into independent monofunctional processes with an allocation factor of 1 for the main flow and allocation factors of 0 for all minor flows. In the example:

$$\mathbf{A'} = \left(\begin{array}{ccc} 1 \times -2 & 100 & 0 \times -2 \\ 10 & 0 & 0 \\ 0 & 0 & 20 \end{array} \right) \tag{3.53}$$

Matrix inversion simply yields

$$\mathbf{s}' = \begin{pmatrix} 100 \\ 2 \\ 0 \end{pmatrix} \tag{3.54}$$

Applying the same allocation factors to the intervention matrix gives

$$\mathbf{B}' = \begin{pmatrix} 1 \times 1 & 10 & 0 \times 1 \\ 1 \times 0.1 & 2 & 0 \times 0.1 \\ 1 \times 0 & -50 & 0 \times 0 \end{pmatrix} \tag{3.55}$$

so that we again find

$$\mathbf{g} = \mathbf{B}'\mathbf{s}' = \begin{pmatrix} 120 \\ 14 \\ -100 \end{pmatrix} \tag{3.56}$$

The surplus method obviously fails when it is one of the minor flows that is involved in the product system.

For this method, it is possible to write down expressions for the final supply vector. As

$$\tilde{\mathbf{f}} = \mathbf{A}\mathbf{s}' \tag{3.57}$$

we have

$$\tilde{\mathbf{f}} = \begin{pmatrix} 0 \\ 1000 \\ 2000 \end{pmatrix} \tag{3.58}$$

and

$$\mathbf{d} = \begin{pmatrix} 0 \\ 0 \\ -2000 \end{pmatrix} \tag{3.59}$$

for the discrepancy vector.

3.2.5 Other methods*

Above, the computational approaches towards the three major methods for dealing with the multifunctionality problem were presented. These are major, in the sense that the first two methods, substitution and partitioning, are often applied and very often discussed in methodological treatises as well as in concrete LCA studies, and that the third method is easily applied, especially in company-internal studies with unimportant minor flows. More methods for dealing with multifunctionality are conceivable, however. This section will discuss some of them briefly:

- dividing a multifunctional process into its constituent monofunctional processes;

- linear programming;

- using the pseudoinverse;

- merging economic flows.

There may even exist more methods.

In the ISO 14041 standard (ISO, 1998), the favoured method to deal with multifunctional systems is by dividing the multifunctional unit process into subprocesses. The idea is that the partitioning method is to divide a unit process in an artificial way, whereas sometimes there may be more knowledge available about the precise mechanisms that govern the process. An example is a factory that produces brown shoes and black shoes. With the partitioning, one would partition the use of materials and energy, and the releases of pollutants over the shoes with some generic factor, perhaps $50 - 50$. If one takes a better look inside the factory hall, one sees two separate production lines: one for the brown shoes and one for the black shoes. Their material and energy requirements and their releases may be recorded in the usual way, and the data collection ends with two really independent unit processes (although there will probably be certain shared aspects related to buildings and management). This is an extremely valuable approach. In the context of this book, however, it is not of interest, because it just means that a refinement of the technology matrix is introduced without any computational concern.

A variation to the surplus method may be approached by means of linear programming. In a linear programming problem, the question is often to produce a set of goods, where the quantity of each of these goods is given as a minimum, and where an additional optimisation criterion is specified, for instance cost minimisation. In the example, a linear programming formulation would be: find scaling factors $\mathbf{s}$ such that the system produces at least 0 litre of fuel, at least 1000 kWh of electricity and at least 0 MJ of heat. In mathematical terms, the balance equation

$$\mathbf{As} = \mathbf{f} \tag{3.60}$$

is changed into

$$\mathbf{As} \geq \mathbf{f} \tag{3.61}$$

Algorithms for solving a linear programming problem have extensively been discussed in the literature of operations research. A frequently used one is the Simplex method, see, *e.g.*, Baumol (1972) and Chiang (1984). Use of the principles of linear programming yields a solution for the scaling vector **s**, and expressions for the final supply vector $\tilde{\mathbf{f}}$, the discrepancy vector **d**, and the inventory vector **g**. Notice that the linear programming solution requires that the vector **f** be specified as flows that must be non-negative. This seems a natural state of affairs, but we will see in Section 3.9.3 that negative components of **f** sometimes make sense as well. It is not clear how the optimisation criterion should be specified.

Still another way of dealing with the problem of multifunctionality goes back to the problem of having an overdetermined system of equations. In parameter estimation, for instance in fitting the coefficients of an equation to data, there is always an overdetermined system of equations. The task is then to find coefficients that make a best fit to the data. A usual criterion is that the sum of squares of deviations of observed and fitted values is minimised. In the case of a linear system of equations, the classical multiple regression model without a constant term is the most appropriate one to discuss. A system of n equations in 2 unknowns

$$\begin{cases} a_{11} \times s_1 + a_{12} \times s_2 = f_1 \\ a_{21} \times s_1 + a_{22} \times s_2 = f_2 \\ \quad \dots \\ a_{n1} \times s_1 + a_{n2} \times s_2 = f_n \end{cases} \tag{3.62}$$

can be written as

$$\mathbf{As} = \mathbf{f} \tag{3.63}$$

This leads to the so-called normal equations:

$$\mathbf{A}^{\mathrm{T}}\mathbf{As} = \mathbf{A}^{\mathrm{T}}\mathbf{f} \tag{3.64}$$

where $\mathbf{A}^{\mathrm{T}}$ is the transpose of $\mathbf{A}$, with a general solution provided by

$$\mathbf{s} = \left(\mathbf{A}^{\mathrm{T}}\mathbf{A}\right)^{-1} \mathbf{A}^{\mathrm{T}}\mathbf{f} \tag{3.65}$$

In the example, we have

$$\mathbf{A} = \begin{pmatrix} -2 & 100 \\ 10 & 0 \\ 20 & 0 \end{pmatrix} \tag{3.66}$$

hence

$$\left(\mathbf{A}^{\mathrm{T}}\mathbf{A}\right)^{-1} \mathbf{A}^{\mathrm{T}} = \begin{pmatrix} 0 & 0.2 & 0.04 \\ 0.01 & 0.0004 & 0.0008 \end{pmatrix} \tag{3.67}$$

and

$$\mathbf{s} = \begin{pmatrix} 20 \\ 0.4 \end{pmatrix} \tag{3.68}$$

One important difference with the previously discussed

$$\mathbf{s} = \mathbf{A}^{-1}\mathbf{f} \tag{3.69}$$

is that the new expression can be applied to overdetermined systems of equations, with a rectangular matrix, whereas the original one only works for square matrices. Another important difference is that the balance equations are not exactly satisfied, but that

$$(\mathbf{As} - \mathbf{f})^{\mathrm{T}} (\mathbf{As} - \mathbf{f}) = \sum_{i=1}^{n} ((\mathbf{As})_i - f_i)^2 \tag{3.70}$$

is minimised. Recalling the discrepancy vector $\mathbf{d}$ from a previous section, we may also interpret this as the square of the length of this vector:

$$\sum_{i=1}^{n} ((\mathbf{As})_i - f_i)^2 = \sum_{i=1}^{n} \left(\tilde{f}_i - f_i \right)^2 = \sum_{i=1}^{n} (d_i)^2 = |\mathbf{d}|^2 \tag{3.71}$$

The vector $\mathbf{d}$, which can here be written as

$$\mathbf{d} = \mathbf{As} - \mathbf{f} \tag{3.72}$$

is sometimes known as the residual vector or the minimum residual.

The sum of squares of residuals d_i obviously can not be negative. In general, it will be larger than zero. Only in the exceptional case that the overdetermined system of equations contains redundant equations can it assume the value 0. If not, the balance equations have not been solved, but a solution has been found such that the balance equations are solved as good as possible. In a normal LCA context, this is not acceptable. For instance, in the example we have

$$\tilde{\mathbf{f}} = \begin{pmatrix} 0 \\ 200 \\ 400 \end{pmatrix} \tag{3.73}$$

so that the discrepancy vector is

$$\mathbf{d} = \begin{pmatrix} 0 \\ -800 \\ 400 \end{pmatrix} \tag{3.74}$$

Moreover, the sum of squares requires that an inner product be defined over the linear space. In our definition of the co-ordinate system, a linear space was introduced, but without an inner product. Evidently, it makes no sense to speak of the length of a vector in a space that is spanned by a basis that represents litre of fuel, kWh of electricity, and MJ of heat. The unit of such a vector would have the non-sensical unit $\sqrt{(\text{litre of fuel})^2 + (\text{kWh of electricity})^2 + (\text{MJ of heat})^2}$. Notice also that the procedure outlined above is dependent upon the choice of the units. That is, a different inventory vector will be found when one changes litre of fuel into gallons of fuel.

A final way of dealing with the multifunctionality problem goes back even more to its origins: the problem of having an overdetermined system of equations. The technology matrix is rectangular, while we want it to be square. We have discussed several approaches to increasing the number of columns, by inserting an avoided unit process or by splitting a multifunctional unit process into several independent unit processes. We also have discussed an approach to decrease the number of rows, by partitioning the matrix into a part that does participate in the balance equations and a part that does not. One additional approach to decreasing the number of rows of the technology matrix is by merging rows of that matrix, *i.e.* by merging economic flows. Suppose that we consider the production of electricity and heat as a plain production of energy. Then we can merge the economic flows electricity and heat into one economic flow, labeled energy. Let us start to choose for expressing energy in MJ and converting 18 MJ of heat into 5 kWh of heat. Then 10 kWh of electricity and 5 kWh of heat can be merged to yield 15 kWh of energy. The technology matrix and the final demand vector then become

$$\mathbf{A}' = \begin{pmatrix} -2 & 100 \\ 15 & 0 \end{pmatrix} \tag{3.75}$$

and

$$\mathbf{f}' = \begin{pmatrix} 0 \\ 1000 \end{pmatrix} \tag{3.76}$$

The technology matrix is square and can be inverted. It gives

$$\mathbf{s}' = \begin{pmatrix} 66.7 \\ 1.33 \end{pmatrix} \tag{3.77}$$

for the scaling vector. This may look fine, but one should not forget that the final demand has implicitly been reformulated in terms of energy instead of electricity. Applying the scaling factors that were just found to the original technology matrix gives

$$\tilde{\mathbf{f}} = \mathbf{As}' = \begin{pmatrix} 0 \\ 667 \\ 1200 \end{pmatrix} \tag{3.78}$$

In words, the 1000 kWh of energy that is supplied consists in fact of 667 kWh of electricity and 1200 MJ (= 333 kWh) of heat. The discrepancy vector is

$$\mathbf{d} = \tilde{\mathbf{f}} - \mathbf{f} = \begin{pmatrix} 0 \\ -333 \\ 1200 \end{pmatrix} \tag{3.79}$$

The method outlined above may seem awkward. Yet, is implicitly applied in many process specifications. For instance, an aeroplane transports different types of passengers (economy class, business class, last minute, children, adult) at the same time, though we often say that it transports a certain number of passengers. And in input-output analysis (see Section 5.1 for more on this) the assumption of homogeneity of production is even indispensable.

3.2.6 Brief discussion

We do not exclude that there may be even more approaches to solving the multifunctionality problem. But, as a matter of fact, the only solutions that are currently applied on a wide scale are the substitution method and the partitioning method. In this book we take a quite neutral position in the debate surrounding the two major methods for dealing with allocation. We place an emphasis on the computational principles behind these methods, rather than running into arguments of the strong and weak points of the methods, and their realm of applicability.

3.3 System boundaries

In inventory analysis, the collection of unit processes included is referred to as the product system. It may be visualised by means of a flow chart. The product system has interactions with the environment by environmental (or elementary) flows. These system-wide environmental flows, the environmental interventions, are the prime object of interest in the inventory analysis. It is clear that it is essential to define for each flow if it is categorised as an economic or as an environmental flow. The environment can be conceived as a system as well, the environmental system, containing many environmental processes related to the fate of substances released to the environment and related to the regeneration of resources extracted from the environment. Therefore, the categorisation of a flow as an economic or an environmental flow is equivalent to the categorisation of a process as an economic or an environmental process. Thus the issue of defining the boundaries of the system and between the two systems (economy and environment) is of interest.

Notice that we have adopted a point of view that differs from that of Clift *et al.* (1998), who distinguish a product system and the environment, where the environment is explicitly not regarded as being part of the system, but is used "in its original (thermodynamic) sense as that *which surrounds the system under study*" (no paging, original italics). One important reason for adopting a different stance is that we disagree with this interpretation of the system-environment dichotomy, and that we think that the term 'environment' is used inappropriately here. In thermodynamics, the environment (or sometimes called surroundings) is assumed to be much larger than the system. It is assumed to be an infinite source and sink of materials and/or energy (see, for instance, the discussion on 'heat reservoirs' by Reif (1964, p.157–159)), so that there are never 'environmental problems.' The environment is not the object of study in thermodynamics, while it forms the main motivation to execute an LCA-study.

Typically, the setting of system boundaries involves the following types of decisions:

- the boundaries between the product system and the environmental system;

- the exclusion of unit processes from the product system because they are unimportant;

- the exclusion of unit processes from the product system because they belong to a different product system.

We will discuss these different boundaries below.

3.3.1 Economic and environmental flows

The first type of boundary – between economic and environmental system – has been discussed in Section 2.1 as the act of partitioning a process matrix **P** into two parts: one with economic flows and one with environmental flows. Economic flows were placed in the technology matrix **A**, environmental flows in the intervention matrix **B**. No guidance has been given on when a flow belongs to the set of economic flows and when to the set of environmental flows. This is beyond the scope of this book, and it is one of the important tasks for guidebooks on LCA. From a computational perspective, the choice of categorisation of flows into economic and environmental is unimportant.

Formally, we may partition the set of flows P into two sets A and B, with the conditions that the partitioning is complete:

$$A \cup B = P \tag{3.80}$$

and non-overlapping:

$$A \cap B = \emptyset \tag{3.81}$$

A general rule for categorising a flow into A or B cannot be given; it depends on the situation and on the investigator's personal view on what is economy and what is environment. A forest can be seen as a production facility, a factory of trees. One downstream activity is logging trees, converting the economic flow trees into a different economic flow: logged trees. It can also be seen as a natural endowment. Then the activity of logging trees converts the environmental flow trees into the economic flow logged trees. At the output-side of the product system, similar ambiguities arise with respect to landfill.

3.3.2   Cut-off

The exclusion of unit processes from the product system because they are unimportant or because the data are not available has in Section 3.1 been referred to as applying cut-off. It again involves the partitioning of a set of flows into two subsets. Now, it is the set of economic flows A which is partitioned into a set A' and a set A". The former set contains all economic flows which do participate in the balance equations $\mathbf{A}'\mathbf{s} = \mathbf{f}'$, while the latter set is excluded from this balancing procedure, leads to $\mathbf{f}'' = \mathbf{A}''\mathbf{s}$ and hence to a non-zero discrepancy vector **d**.

Again, criteria for applying cut-off can be found in many guidebooks for LCA, and the computational details are independent from the choice of cut-off rules.

It is possible to make cut-off an 'automatic' procedure, from a computational point of view at least. Cut-off is needed when no process is available for supplying a good or treating a waste. In Section 3.1, for instance, no processes for producing new generators or treating used generators were known. In the technology matrix, this can for the third row be detected from the fact that there is a row (the third one, denoting new generators) for which certain processes have a negative entry (here the first process) and for which no entry is positive. In that case, an algorithm may partition the technology matrix so that the third flow is transferred to $\mathbf{A}''$. A similar argument holds for the fourth flow, but here certain processes have a positive entry, but no process with a negative entry is available. A formal criterion of automatic cut-off cannot yet be defined here; this will take place in Section 3.3.4.

In addition to automatic cut-off, cut-off may also be made manually, even when it is not needed, for instance to study what the influence of not having the data would have been. Take the example with

$$\mathbf{A} = \begin{pmatrix} -2 & 100 \\ 10 & 0 \end{pmatrix} \qquad (3.82)$$

and suppose that we wish to study what the effect would be of cutting-off fuel production. This would reduce the technology matrix to

$$\mathbf{A}' = \begin{pmatrix} 10 & 0 \end{pmatrix} \qquad (3.83)$$

which can of course not be inverted. To study the influence of cut-off, we must also remove the process that has been cut-off (the second column), or we must employ the pseudoinverse (see Section 3.5.2) to find a solution.

3.3.3 Other product systems

Section 3.2 discusses how processes that deliver more than one useful output may be treated in the matrix formalism. A connection to the setting of system boundaries needs to be discussed as well. When chlorine is needed in the product system, the technology matrix may contain a process that produces chlorine and caustic soda. The downstream processes that use the chlorine are naturally included in the system, but the downstream processes that use the caustic soda are not: they are considered to belong to a different product system. As a matter of fact, they may be included in the system, and they will receive a zero scaling factor, which means that they play no role and could have been left out of the system as well (see also Section 3.8.3). The economy of research will in many cases lead to sparsity in data collection, so that no data for such processes will be collected and they will indeed be left out.

From a computational perspective, we may act like in the case of cut-off. The 'superfluous' co-product, caustic soda in the example, will be transferred to A", while the 'required' co-product, chlorine, enters A'. Computation then proceeds as before on the basis of $\mathbf{A's} = \mathbf{f'}$. Of course, this approach works for the partitioning method (Section 3.2.3) and for the surplus method (Section 3.2.4). For the substitution method (Section 3.2.2) things are different, because the 'superfluous' co-product does participate in the balancing procedure and so defines the scaling factor for the 'avoided' process. For the downstream processes that use the caustic soda, a similar situation holds: they may be excluded, or they may be included and will receive a scaling factor 0. Only for the substitution method difficulties will arise: the downstream process that uses the caustic soda and the stand-alone (avoided) process of caustic soda production will be interpreted as competitors, producing the same economic flow. Section 3.4.4 provides a solution by distinguishing 'brands' of caustic soda.

3.3.4 Goods and wastes

The surplus method can be seen as conceptually identical to cut-off. In both cases, the rectangularity of the technology matrix is remedied by shifting rows to a matrix $\mathbf{A}''$ that does not partake in the balancing procedure. Despite this conceptual similarity, they are not factual identical. This section discusses the distinguishing feature in detail, even though it is not strictly a matter of system boundaries.

In Section 3.1.1, a hypothetical process was introduced with the co-ordinates

$$\mathbf{p} = \begin{pmatrix} -2 \\ 10 \\ -10^{-12} \\ 10^{-12} \end{pmatrix} \tag{3.84}$$

in a basis, defined as

$$\begin{pmatrix} \text{litre of fuel} \\ \text{kWh of electricity} \\ \text{new generator} \\ \text{used generator} \end{pmatrix} \tag{3.85}$$

This may be contrasted with the hypothetical process in Section 3.2.1 with co-ordinates

$$\mathbf{p} = \begin{pmatrix} -2 \\ 10 \\ 18 \end{pmatrix} \tag{3.86}$$

in the basis

$$\begin{pmatrix} \text{litre of fuel} \\ \text{kWh of electricity} \\ \text{MJ of heat} \end{pmatrix} \tag{3.87}$$

We see that both processes have two outputs: electricity and used generators in the first case, and electricity and heat in the second case. In both cases, the system will not include a process that absorbs the second output, *viz.* used generators and heat. Still, we adopt a different approach to dealing with these processes: the first one is considered as producing a waste flow that is subject to a cut-off procedure, while the second one is considered as a multifunctional process and may be subject to an allocation procedure. This section is devoted to an analysis of this difference.

There is a clear difference in meaning between producing used generators and producing heat. The first one is a waste, in the sense that it is not sold to a next process, but that the next process is paid for receiving it. For a human investigator, the difference between used generators and heat may be clear. For an algorithm it is not clear. Moreover, there are situations in which the distinction between a valuable good and a waste that is to be treated is far from clear. For instance, many low-quality flows from industrial processes may be seen as a low-value co-product or as a waste, depending on the capacities for upgrading. Although wastes – in the sense of goods with a negative value – are sometimes regarded as co-products, and a process producing one valuable good and one or more wastes as a co-producing product (Baumgärtner, 2000), we take a more strict stance here. Goods have a positive economic value, wastes have a negative economic value. Thus a product producing one good and one waste is a monofunctional process, and no allocation procedure is required. To emphasise the fact that wastes are not goods, the term 'bads' is sometimes used for them (Lipsey & Steiner (1978, p.6), Heijungs (1997)). Anyhow, be it for computational or mnemonical reasons, it is clear that a label must be attached to every economic flow, indicating if it is a good or a waste.

Formally, the set of economic flows A must be partitioned into two sets: the goods G and the wastes W. The economic part of an arbitrary process vector will then be written as

$$\mathbf{a} = \left(\frac{\mathbf{g}}{\mathbf{w}} \right) \tag{3.88}$$

With this, the criterion for multifunctionality then becomes as follows:

$$\exists i \neq k : ((g_i > 0 \wedge g_k > 0) \vee (w_i < 0 \wedge w_k < 0) \vee (g_i > 0 \wedge w_k < 0)) \tag{3.89}$$

or in words: two (or more) outputs of goods, two (or more) inputs of wastes, or one (or more) output of goods and one (or more) input of wastes. The three cases correspond to those mentioned in Section 3.2:

$$\exists i \neq k : (g_i > 0 \wedge g_k > 0) \tag{3.90}$$

for co-production,

$$\exists i \neq k : (w_i < 0 \wedge w_k < 0) \tag{3.91}$$

for combined waste treatment, and

$$\exists i \neq k : (g_i > 0 \wedge w_k < 0) \tag{3.92}$$

for recycling.

The criterion of automatic cut-off is defined at the level of the system, not of the individual process. It thus requires consideration of the technology matrix:

$$\exists i : \quad (((\exists j : g_{ij} < 0) \wedge (\nexists k : g_{ik} > 0)) \vee \\ ((\exists j : w_{ij} > 0) \wedge (\nexists k : w_{ik} < 0))) \tag{3.93}$$

or in words: a good i is cut-off when it is the input of one (or more) process j and not the output of any process, and a waste i is cut-off when it is the output of one (or more) process j and not the input of any process.

3.4 Choice of suppliers

It is often the case that one economic flow is produced by more than one unit process with different characteristics. For instance, electricity may be produced by fuel-driven generators or by coal-fired plants. When this happens, the computational procedure outlined in Chapter 2 can be followed, but only when certain rules are kept in mind.

3.4.1 Formulation of the problem

Let us change the basis of the economic part into

$$\begin{pmatrix} \text{litre of fuel} \\ \text{kWh of electricity} \\ \text{kg of coal} \end{pmatrix} \tag{3.94}$$

and include two new unit processes: production of electricity by a coal-fired plant and mining of coal. The technology matrix becomes

$$\mathbf{A} = \begin{pmatrix} -2 & 100 & 0 & 0 \\ 10 & 0 & 10 & 0 \\ 0 & 0 & -5 & 50 \end{pmatrix} \tag{3.95}$$

With a final demand vector

$$\mathbf{f} = \begin{pmatrix} 0 \\ 1000 \\ 0 \end{pmatrix} \tag{3.96}$$

the inventory problem becomes to solve

$$\mathbf{As} = \mathbf{f} \tag{3.97}$$

for the scaling factors $\mathbf{s}$. This problem cannot be solved with matrix inversion, because the technology matrix $\mathbf{A}$ is not square. As in the case of multifunctionality, it is a rectangular matrix, but this time with more columns than rows.

If we expand the matrix equation as a system of linear equations, we find

$$\begin{cases} -2 \times s_1 + 100 \times s_2 + 0 \times s_3 + 0 \times s_4 = 0 \\ 10 \times s_1 + 0 \times s_2 + 10 \times s_3 + 0 \times s_4 = 1000 \\ 0 \times s_1 + 0 \times s_2 + -5 \times s_3 + 50 \times s_4 = 0 \end{cases} \tag{3.98}$$

The solution to the original system $\mathbf{s} = \begin{pmatrix} 2 & 100 & 0 & 0 \end{pmatrix}^{\mathrm{T}}$ works, as does $\mathbf{s} = \begin{pmatrix} 0 & 0 & 100 & 10 \end{pmatrix}^{\mathrm{T}}$ and $\mathbf{s} = \begin{pmatrix} 1 & 50 & 50 & 5 \end{pmatrix}^{\mathrm{T}}$. In fact, there is an infinite number of solutions. In the case of multifunctionality, the system was overdetermined; here it is underdetermined, as there are fewer equations than unknowns.

To solve the problem of indeterminacy, two approaches will be discussed. These two approaches are not alternatives, as was the case for multifunctionality. Rather, they provide two different views on the meaning of a choice of suppliers in the context of LCA.

3.4.2 Comparison of alternative systems

The first approach is geared towards considering the alternative ways of producing a certain product as one of main qualities of LCA. Such a comparison necessarily involves making the products distinguishable by their origin. For instance, we change the basis into

$$\begin{pmatrix} \text{litre of fuel} \\ \text{kWh of fuel-derived electricity} \\ \text{kg of coal} \\ \text{kWh of coal-derived electricity} \end{pmatrix} \tag{3.99}$$

The technology matrix is the changed into

$$A = \begin{pmatrix} -2 & 100 & 0 & 0 \\ 10 & 0 & 0 & 0 \\ 0 & 0 & -5 & 50 \\ 0 & 0 & 10 & 0 \end{pmatrix} \qquad (3.100)$$

This is a square, invertible matrix. But what to do with the final demand vector? We introduce two of them, one to represent a reference flow of 1000 kWh of fuel-derived electricity and one to represent a reference flow of 1000 kWh of coal-derived electricity. Thus

$$\mathbf{f}_1 = \begin{pmatrix} 0 \\ 1000 \\ 0 \\ 0 \end{pmatrix} \text{ and } \mathbf{f}_2 = \begin{pmatrix} 0 \\ 0 \\ 0 \\ 1000 \end{pmatrix} \qquad (3.101)$$

This then yields two scaling vectors:

$$\mathbf{s}_1 = \begin{pmatrix} 100 \\ 2 \\ 0 \\ 0 \end{pmatrix} \text{ and } \mathbf{s}_2 = \begin{pmatrix} 0 \\ 0 \\ 100 \\ 10 \end{pmatrix} \qquad (3.102)$$

with direct consequences for two inventory vectors $\mathbf{g}_1$ and $\mathbf{g}_2$.

The issue of comparing several product alternatives is discussed more extensively in Section 3.8.1.

3.4.3   Representing a mixing process

The second approach starts from observing that many products on the market are bought from a large pool of identical products, where many producers deliver only a certain share of the total volume available. For instance, electricity taken from the grid is a mix of electricity by hundreds of power plants, some old, some new, some hydro-operated, some nuclear, etc. All these electricity flows are mixed, and one can point out at a certain time, or averaged over one year, the proportions that all individual electric power plants contribute to the total volume. The task is to capture this idea in the matrix formalism. Notice that this clearly demonstrates that the technology matrix, that contains the process data, is based on more then technological information alone. In this case, market-based aspects, namely market shares, are included. In the case of consumption processes, behavioural aspects are included as well.

In the example, let us assume that the share of fuel-derived electricity is 60% and that of coal-derived electricity 40%. Then we define a mixing process that has an output of 1 kWh of mixed electricity and inputs of 0.6 kWh of fuel-derived electricity and 0.4 kWh of coal-derived electricity. In matrix form, relative to a basis

$$\begin{pmatrix} \text{litre of fuel} \\ \text{kWh of fuel-derived electricity} \\ \text{kg of coal} \\ \text{kWh of coal-derived electricity} \\ \text{kWh of mixed electricity} \end{pmatrix} \tag{3.103}$$

the technology matrix is

$$\mathbf{A} = \begin{pmatrix} -2 & 100 & 0 & 0 & 0 \\ 10 & 0 & 0 & 0 & -0.6 \\ 0 & 0 & -5 & 50 & 0 \\ 0 & 0 & 10 & 0 & -0.4 \\ 0 & 0 & 0 & 0 & 1 \end{pmatrix} \tag{3.104}$$

and the final demand vector

$$\mathbf{f} = \begin{pmatrix} 0 \\ 0 \\ 0 \\ 0 \\ 1000 \end{pmatrix} \tag{3.105}$$

This yields a scaling vector

$$\mathbf{s} = \begin{pmatrix} 60 \\ 1.2 \\ 40 \\ 4 \\ 1000 \end{pmatrix} \tag{3.106}$$

with an inventory vector that is also a weighted mix of those for the individual modes of electricity generation.

The example was about mixing of electricity. Electricity is indeed a fairly homogeneous product which can be considered to be mixed. Other cases of mixing include bulk materials, like steel and oil. But from a macro-point of view, small distinctions between competitive products may be ignored in certain cases. For instance, when comparing the environmental impacts of coke versus beer, one could decide to not compare coke of brand X with beer of brand Y, but to compare the weighted mix of all coke brands with the weighted mix of all beer brands. A still higher level of mixing is required when one compares soft drinks with alcoholic drinks, which would involve the mixing of coke, lemon juice, etc, as well as the mixing of beer, wine, etc. One could also compare tomatoes from organic agriculture with those from greenhouse culture. This would involve mixing of all organically grown tomatoes and the mixing of all tomatoes from greenhouses. Choices like these are necessary for LCAs at a high levels of aggregation, and represent cases in which LCA is used to analyse consumption scenarios.

3.4.4 Brief discussion

A general lesson from the above discussion is that every product should come in one brand only. There should not be two unit processes producing the same product. If there are, the two outputs should be made distinguishable by reserving a separate row for each. Appropriate names may then be conceived, like fuel-derived versus coal-derived electricity or coke of brand X versus brand Y. It may happen that technologies change in the course of time. For instance, electricity production during construction of a house may be different from electricity production during demolition of that house, a hundred years later. Naming of economic flows could then include a temporal qualification, like electricity AD 2000 and electricity AD 2100. Similarly, spatial qualifications may be added to distinguish Swiss electricity from Austrian electricity. Sections 9.2 and 9.3 provide a further discussions on spatial and temporal qualifications.

One consequence of the statement that every product should come in one brand only, is that a technology matrix that includes a multifunctional process cannot be square. Thus, the matrix (3.23) presented to illustrate the substitution method (Section 3.2.2), namely

$$\mathbf{A}' = \begin{pmatrix} -2 & 100 & -5 \\ 10 & 0 & 0 \\ 18 & 0 & 90 \end{pmatrix} \tag{3.107}$$

is ill-constructed. The proper representation should be the one presented later onwards in (3.28), namely

$$\mathbf{A}' = \begin{pmatrix} -2 & 100 & -5 \\ 10 & 0 & 0 \\ 18 & 0 & 0 \\ 0 & 0 & 90 \end{pmatrix} \tag{3.108}$$

where the third row represents MJ of heat from facility X, and the fourth row MJ of heat from facility Y. The crucial step in the substitution method is that an explicit statement is made regarding the indistinguishable features of the heat from brand X and brand Y. This assumed equivalency then enables one to merge the two rows, so that indeed

$$\mathbf{A}' = \begin{pmatrix} -2 & 100 & -5 \\ 10 & 0 & 0 \\ 18 & 0 & 90 \end{pmatrix} \tag{3.109}$$

is obtained, where the third row represents MJ of heat from facility X or Y. Observe that, contrary to the situation in Section 3.4.3, there is no information on market shares of brands X and Y included in the above technology matrix.

The 'one brand axiom' is not adhered to in many treatments of LCA; see also Section 3.2.2 for an example. In Section 4.5, we will see how economic models, most notable the supply/use framework and the activity analysis are complicated by this defect.

3.5 Closed-loop recycling

In LCA, recycling is defined as the situation in which a unit process transforms a negatively valued product or material (*i.e.* a waste) into a positively valued product or material. The situation that a process transforms a product or material with a small positive value into a product or material with a higher value is not regarded as closed-loop recycling, but as a normal production process, like production of steel plates from bulk steel. In general, a recycling process is one of the many unit processes, and representation of a recycling process is as usual. We have already discussed recycling as one of the examples of a multifunctional process (see Section 3.2), and for handling these, many methods have been described. So there does not seem to be much need for a separate discussion of recycling. Closed-loop recycling, however, presents a special case.

3.5.1   Formulation of the problem

We speak of closed-loop recycling when secondary material produced by a recycling process is completely fed back into one of the unit processes of the same product system. When the material is transferred to another product system, we speak of open-loop recycling. Open-loop recycling can be treated in the way described in Section 3.2. Finally, when a recycling process delivers only a part of the secondary material to the product system itself, and part to another product system, we have a case of a mixture of closed- and open-loop recycling. This section will concentrate on closed-loop recycling.

Let us, for instance, suppose that a waste stream from fuel production can be used in the electricity production process. We might call this recycling of waste into electricity, or waste incineration with energy recovery. Anyhow, consider the following technology matrix:

$$\mathbf{A} = \begin{pmatrix} -2 & 100 \\ 10 & 0 \\ -1 & 50 \end{pmatrix} \tag{3.110}$$

where the basis of the economic part is as follows:

$$\begin{pmatrix} \text{litre of fuel} \\ \text{kWh of electricity} \\ \text{kg of waste} \end{pmatrix} \tag{3.111}$$

The technology matrix is not square, and can not be inverted. This means that the computational procedure described in Chapter 2 is not directly applicable when a closed-loop recycling process is present. One may of course apply an allocation step, either according to the substitution method (Section 3.2.2) or according to the partitioning method (Section 3.2.3). But, there seems to be a broad consensus that in case of closed-loop recycling, "the need for allocation is avoided since the use of secondary materials displaces the use of virgin (primary) materials" (ISO, 1998, p.12).

We can see that allocation is not needed by studying the system of balance equations:

$$\begin{cases} -2 \times s_1 + 100 \times s_2 = 0 \\ 10 \times s_1 + 0 \times s_2 = 1000 \\ -1 \times s_1 + 50 \times s_2 = 0 \end{cases} \tag{3.112}$$

It is immediately clear that the first and the third balance equation express the same: the first one is the third one multiplied by 2. The third equation (or the first one, for that matter) is redundant. Intuitively, we can figure out the solution to be

$$\mathbf{s} = \begin{pmatrix} s_1 \\ s_2 \end{pmatrix} = \begin{pmatrix} 100 \\ 2 \end{pmatrix} \tag{3.113}$$

The problem is now: how can we, in a formal setting, find a solution? Heijungs & Frischknecht (1998) propose to replace the matrix inverse by the pseudoinverse; see hereafter for a presentation. Two further sections show what happens when an allocation step is performed anyhow: the partitioning method (3.5.3) and the substitution methods (3.5.4).

3.5.2 Solution with the pseudoinverse

We return briefly to the discussion of regression analysis in Section 3.2.5. In regression analysis, an overdetermined system is solved such that an approximation appears that is as close as possible to the question. We interpreted the length of the discrepancy vector as a variable to be minimised. The value obtained for this length is a measure of the quality of the fit. Now, suppose that we find a length 0. This is the smallest length that can be obtained, and it corresponds to the case of a perfect fit. This means that application of the regression framework offers an approach to solving the redundant system of equations in the case of closed-loop recycling. Indeed, the formerly discussed expression

$$\mathbf{s} = \left(\mathbf{A}^{\mathrm{T}} \mathbf{A} \right)^{-1} \mathbf{A}^{\mathrm{T}} \mathbf{f} \tag{3.114}$$

gives in this case a perfect answer:

$$\mathbf{s} = \begin{pmatrix} 0 & 0.1 & 0 \\ 0.008 & 0.002 & 0.004 \end{pmatrix} \begin{pmatrix} 0 \\ 1000 \\ 0 \end{pmatrix} = \begin{pmatrix} 100 \\ 2 \end{pmatrix} \tag{3.115}$$

Moreover, as the final supply vector in this case coincides with the final demand vector, the discrepancy vector is a null-vector:

$$\mathbf{d} = \mathbf{0} \tag{3.116}$$

In this case, we have thus managed to find a perfect solution in an overdetermined system with redundant equations. The expression

$$\mathbf{s} = \left(\mathbf{A}^{\mathrm{T}} \mathbf{A} \right)^{-1} \mathbf{A}^{\mathrm{T}} \mathbf{f} \tag{3.117}$$

may be rewritten as

$$s = \mathbf{A}^+ \mathbf{f} \tag{3.118}$$

where

$$\mathbf{A}^+ = \left(\mathbf{A}^T \mathbf{A}\right)^{-1} \mathbf{A}^T \tag{3.119}$$

Under certain conditions (see Harville (1997, p.495) and Albert (1972)), the matrix $\mathbf{A}^+$ is identical to a matrix that is known as the Moore–Penrose inverse or the pseudoinverse. Notice, however, that the term pseudoinverse is sometimes used to indicate the broader class of generalised inverses. We thus see that the inventory problem

$$\mathbf{A}\mathbf{s} = \mathbf{f} \tag{3.120}$$

with a rectangular matrix due to closed-loop recycling may be solved using the pseudoinverse $\mathbf{A}^+$ with the equation

$$\mathbf{s} = \mathbf{A}^+ \mathbf{f} \tag{3.121}$$

with a perfect match between final demand and final supply:

$$\mathbf{d} = \tilde{\mathbf{f}} - \mathbf{f} = \mathbf{0} \tag{3.122}$$

An allocation procedure may thus indeed be omitted for multifunctional processes that are involved in closed-loop recycling, and the pseudoinverse provides an exact solution in such cases. One should note that in real-world calculations $\mathbf{d}$ may deviate from $\mathbf{0}$ due to computational round-off (see Section 6.6).

The pseudoinverse of a square invertible matrix is equal to the normal inverse. This means that the pseuodinverse can in principle always be used, even in the case of a square invertible matrix. It is interesting, however, to investigate the results if the multifunctional process would have been treated with one of the methods that has been discussed in Section 3.2. We will discuss what happens when the partitioning method is used and when the substitution method is used in two separate sections.

3.5.3   Comparison with the partitioning method*

Suppose that we partition the multifunctional process that treats waste and produces electricity into two independent monofunctional processes. Concentrating on the technology matrix, we must introduce two allocation factors, λ_1 and λ_2, to partition the input of fuel among the two independent processes:

$$\mathbf{A'} = \begin{pmatrix} -2 & 100 & 0 \\ \lambda_1 \times 10 & 0 & \lambda_2 \times 10 \\ 0 & 50 & -1 \end{pmatrix} \tag{3.123}$$

This is a square matrix, of which the inverse can be shown to be

$$\mathbf{A'}^{-1} = \frac{1}{\lambda_1 + \lambda_2} \begin{pmatrix} -\frac{1}{2}\lambda_1 & \frac{1}{10} & \lambda_2 \\ \frac{1}{100}\lambda_1 & \frac{1}{500} & \frac{1}{50}\lambda_2 \\ \frac{1}{2}\lambda_1 & \frac{1}{10} & -\lambda_1 \end{pmatrix} \tag{3.124}$$

provided that $\lambda_1 + \lambda_2 \neq 0$. Multiplication with the final demand vector yields

$$\mathbf{s'} = \frac{1}{\lambda_1 + \lambda_2} \begin{pmatrix} 100 \\ 2 \\ 100 \end{pmatrix} \tag{3.125}$$

Upon the usual restriction that $\lambda_1 + \lambda_2 = 1$ this reduces to

$$\mathbf{s'} = \begin{pmatrix} 100 \\ 2 \\ 100 \end{pmatrix} \tag{3.126}$$

Observe that the two processes that were obtained through partitioning have the same scaling factor, 100. This means that, although they may be varied independently, they are actually used to the same extent. Their independence is not made use of. We moreover see that the same scaling factors as in the unpartitioned case are found, 100 and 2, upon the restriction that the sum of the partitioned processes adds up to the original multifunctional process. We may interpret this as that one could have partitioned the multifunctional process according to any allocation principle, mass, energy, costs, but that in the case of closed-loop recycling this choice of allocation principle is immaterial as long as the 100%-rule is satisfied. Notice, however, that we have not provided a general proof of this.

3.5.4 Comparison with the substitution method*

As a second case, let us examine how the substitution method works in the case of closed-loop recycling. Assume that an avoided process is added to the system, to account for the avoided treatment of waste by using it for electricity generation. The specification of this avoided process is left open for the moment, we assume that it requires κ_1 litre of fuel and κ_2 kWh of electricity to absorb 1 kg of waste. Hence, the technology matrix becomes

$$\mathbf{A}' = \begin{pmatrix} -2 & 100 & -\kappa_1 \\ 10 & 0 & -\kappa_2 \\ -1 & 50 & -1 \end{pmatrix} \tag{3.127}$$

The inverse of this matrix is

$$\mathbf{A}'^{-1} = \frac{1}{\kappa_1 - 2} \begin{pmatrix} -\frac{1}{10}\kappa_2 & \frac{1}{10}(\kappa_1 - 2) & \frac{1}{5}\kappa_2 \\ -\frac{1}{500}(10 + \kappa_2) & \frac{1}{500}(\kappa_1 - 2) & \frac{1}{250}(5\kappa_1 + \kappa_2) \\ -1 & 0 & 2 \end{pmatrix} \tag{3.128}$$

Multiplication with the final demand vector gives

$$\mathbf{s}' = \begin{pmatrix} 100 \\ 2 \\ 0 \end{pmatrix} \tag{3.129}$$

provided that

$$\kappa_1 - 2 \neq 0 \tag{3.130}$$

This last requirement comes down to demanding that the technology matrix is non-singular, in this case that the avoided process is not a linear combination of the other processes.

It indeed seems obvious to require that a zero scaling factor is obtained for the avoided process in case of closed-loop recycling. After all, this process is avoided when its function is substituted by an external co-product flow, but in the case of closed-loop recycling the co-product remains system-internal.

3.5.5   Brief discussion

The third approach towards dealing with multifunctional processes that was discussed in Section 3.2 was the surplus method. This was shown to be mathematically equivalent to an extreme form of partitioning. And above, we have seen that allocation factors, even extreme ones, do not matter in treating closed-loop recycling with partitioning as long as the 100%-rule is not violated.

The presentation of closed-loop recycling is important for several reasons:

- it shows how the matrix approach can deal with closed-loop recycling without allocating;

- it shows that use of the partitioning method with an arbitrary basis for the allocation factors is equivalent to the approach without allocation (as long as the 100%-rule is kept in mind);

- it shows that use of the substitution method is equivalent to the approach without allocation (as long as no singular technology matrix is obtained);

- it shows that, when an allocation method is used to treat systems with closed-loop recycling, the technology matrix can be assumed to be square.

This latter fact implies that the rest of this book can be restricted to a discussion of the normal inverse of a square matrix, and that there is thus no need to generalise the discussion to the pseudoinverse of a rectangular matrix.

3.6 Inclusion of aggregated systems*

It frequently happens that one has performed the LCA-technique to calculate the system-wide interventions of a certain product, and that one wishes to use these system-wide results in a different LCA-study. For instance, having calculated the interventions associated with 1 kWh of electricity from a large database of unit processes, one wants to use these results as one module in the calculations for a refrigerator. Indeed, the database by Frischknecht *et al.* (1993) is intended for such uses. The inclusion of such an aggregated system deserves a brief discussion. It should be made clear that this section discusses the incorporation of system wide (*i.e.*, cradle-to-grave, cradle-to-gate, or gate-to-grave) results that are obtained from an LCA, not from an input-output analysis. That topic finds a place in Section 5.4 under the name hybrid analysis.

We start with the example system of Section 2.2, but we already reserve a position for the economic flows with a third dimension, representing 1 hr of cooling. The aggregated system, delivering 1000 kWh of electricity is then

$$
\mathbf{q} = \begin{pmatrix} \mathbf{f} \\ \mathbf{g} \end{pmatrix} = \begin{pmatrix} 0 \\ 1000 \\ 0 \\ \overline{120} \\ 14 \\ -100 \end{pmatrix}
\tag{3.131}
$$

We have written it as $\mathbf{q}$, because it is composed of a final demand vector $\mathbf{f}$ and an inventory vector $\mathbf{g}$. Now, we will consider it as one of the vectors of processes which builds such a system. This means that we form a new basis

$$\begin{pmatrix} \text{kWh of electricity} \\ \text{hr of cooling} \\ \overline{\text{kg of carbon dioxide}} \\ \text{kg of sulphur dioxide} \\ \text{litre of crude oil} \end{pmatrix} \tag{3.132}$$

and that the aggregated system is one of the processes that builds the system in this new space:

$$\mathbf{p_1} = \begin{pmatrix} 1000 \\ 0 \\ \overline{120} \\ 14 \\ -100 \end{pmatrix} \tag{3.133}$$

The process of using a refrigerator, neglecting the depreciation of the refrigerator itself, is the second process. For instance, it could be

$$\mathbf{p_2} = \begin{pmatrix} -10 \\ 24 \\ \overline{0} \\ 0 \\ 0 \end{pmatrix} \tag{3.134}$$

indicating that 24 hr of cooling requires 10 kWh of electricity and is itself free of environmental interventions. Then, the technology matrix is

$$\mathbf{A} = \begin{pmatrix} 1000 & -10 \\ 0 & 24 \end{pmatrix} \tag{3.135}$$

and the intervention matrix

$$\mathbf{B} = \begin{pmatrix} 120 & 0 \\ 14 & 0 \\ -100 & 0 \end{pmatrix} \tag{3.136}$$

Let us impose a reference flow of 100 hr of cooling. The system of equations then becomes

$$\begin{pmatrix} 1000 & -10 \\ 0 & 24 \end{pmatrix} \begin{pmatrix} s_1 \\ s_2 \end{pmatrix} = \begin{pmatrix} 0 \\ 100 \end{pmatrix} \tag{3.137}$$

which can be readily solved for **s**, and hence for **g**.

Notice that we have carried out a reduction in both the number of rows and the number columns of the technology matrix. In the original basis, the system would have a technology matrix like

$$\mathbf{A} = \begin{pmatrix} -2 & 100 & 0 \\ 10 & 0 & -10 \\ 0 & 0 & 24 \end{pmatrix} \tag{3.138}$$

Obviously, the system of equations

$$\begin{pmatrix} -2 & 100 & 0 \\ 10 & 0 & -10 \\ 00 & & 24 \end{pmatrix} \begin{pmatrix} s_1 \\ s_2 \\ s_3 \end{pmatrix} = \begin{pmatrix} 0 \\ 0 \\ 100 \end{pmatrix} \tag{3.139}$$

would lead to different scaling factors **s**, but to the same interventions **g**. The gain in reducing a matrix of dimension 3×3 to one of dimension 2×2 is small, but it will be clear that condensing a 500×500 system to a 1×1 system can be really important, even in an era in which a computer's memory size and clock speed are usually not limiting. A disadvantage of the aggregation is that a breakdown of the results (see the contribution analysis in Section 8.2.1) and an extensive analysis of uncertainties (see Chapter 6 and Section 8.2.4) is no longer possible.

Also notice that one must choose between either including an aggregated system and reducing the system, or not including the aggregated system and not reducing the system. The reason can be illustrated as follows. Suppose that we form a system with processes representing production of electricity, production of fuel, aggregated production of electricity, and using a refrigerator. The process matrix is

$$\mathbf{P} = \begin{pmatrix} \mathbf{A} \\ \overline{\mathbf{B}} \end{pmatrix} = \begin{pmatrix} -2 & 100 & 0 & 0 \\ 10 & 0 & 1000 & -10 \\ 0 & 0 & 0 & 24 \\ \hline 1 & 10 & 120 & 0 \\ 0.1 & 2 & 14 & 0 \\ 0 & -50 & -100 & 0 \end{pmatrix} \tag{3.140}$$

The technology matrix **A** is of dimension 3×4, *i.e.* non-square. At the same time, of course, the columns in **P** are dependent: $\mathbf{p}_3$ is by definition a linear combination of $\mathbf{p}_1$ and $\mathbf{p}_2$. And the second flow, electricity, comes in two brands, from $\mathbf{p}_1$ and from $\mathbf{p}_2$. Normally, a dependency of the columns would lead to a singular matrix, with subsequent problems in inversion. Now, the three types of problems, rectangularity, dependency and violation of the one brand axiom, can be said to cancel one another. Two solutions provide approaches to obtaining an answer. In the first approach, we introduce one new economic dimension: kWh of aggregated energy. The technology matrix becomes

$$\mathbf{A} = \begin{pmatrix} -2 & 100 & 0 & 0 \\ 10 & 0 & 0 & -10 \\ 0 & 0 & 0 & 24 \\ 0 & 0 & 1000 & 0 \end{pmatrix} \tag{3.141}$$

or

$$\mathbf{A} = \begin{pmatrix} -2 & 100 & 0 & 0 \\ 10 & 0 & 0 & 0 \\ 0 & 0 & 0 & 24 \\ 0 & 0 & 1000 & -10 \end{pmatrix} \tag{3.142}$$

depending on the question whether one decides to use the individual processes or the aggregated system for the supply of electricity to the refrigerator. In the first case, the aggregated system is left unused and the computation is open to more types of analysis (contribution analysis, uncertainty analysis, etc.), in the second case, the constituting processes are left unused. For the second solution, we refer to the use of the pseudoinverse (see Section 3.5.2) for solving over- and underdetermined systems of equations.

In any case, the very reason of including aggregated systems in a different system was the reduction of memory requirements and computation time. With that in mind, it seems strange to include an aggregated system along with its constituent processes in one large system.

3.7   Some special unit processes and flows

One of the problems of the matrix formalism is that it is deceptively simple in some respects. This applies in particular to the problems of including certain unit processes in the representation by linear spaces, discussed in Section 2.1. We will discuss consumption processes, transport processes, storage processes and waste treatment processes. We will also be in a position to clarify the meaning of the reference flow, at least in relation to the matrix representation. Finally, we will address the issue of representing waste flows.

3.7.1 Consumption processes

The notion of a consumptive process is one that has been challenged a number of times; see Heijungs (1997, p.25) for a number of quotations. In the present context, one should acknowledge that consuming beer is in fact converting beer into urine, and deriving a certain service or utility, the joy of having a beer, from it. Similarly, consuming a car means converting a new car into a used car and deriving a service, 50,000 kilometres of driving, from it. Next, both material flows (beer and urine, new car and used car) and non-material flows (the joy of having a beer, $50,000$ kilometres of driving) must be represented in the linear space that is assumed to host the unit processes describing such consumptive processes. One might be tempted to leave out the non-material flows, but then, consumption would be a process that converts a valuable good (beer, new cars) into negatively valued waste (urine, used cars). This paradox of value-lowering activities can be resolved by including the increase of welfare due to the deduction of utility from the material flows.

A concrete example may illustrate the representation of consumption processes. Consider a linear space with basis

$$
\begin{pmatrix}
\text{litre of fuel} \\
\text{new cars} \\
\text{used cars} \\
\text{km of car-driving} \\
\hline
\text{kg of carbon dioxide}
\end{pmatrix}
\tag{3.143}
$$

a unit process could be

$$
\mathbf{p} =
\begin{pmatrix}
-5 \\
-10^{-9} \\
10^{-9} \\
100 \\
\hline
10
\end{pmatrix}
\tag{3.144}
$$

to represent the input of fuel, the emission of carbon dioxide and the depreciation of a car involved in driving 100 km by car.

3.7.2 The reference flow

Incorporation of the service-delivering flow of the consumption process is crucial to scale this process in relation to the reference flow. The consumption process is a unit process like all others. This means that is part of the flow chart of processes, and that the inventory problem should address the question of finding scaling factors for this process as well. When the process specification is defined as in Section 3.7.1, and the reference flow is 1000 km of car-driving, we would have a scaling factor of the consumption process of 10.

The reference flow can be seen as one well-defined mode of realising the functional unit. If the functional unit is phrased in terms of 'driving 1000 km,' and if two alternatives, car and train, are chosen as feasible alternatives, we are led to consider the reference flows '1000 km of car-driving' and '1000 km of train-transport'. Suppose that we extend the basis for above into

$$
\begin{pmatrix}
\text{litre of fuel} \\
\text{new cars} \\
\text{used cars} \\
\text{km of car-driving} \\
\text{kWh of electricity} \\
\text{new trains} \\
\text{used trains} \\
\underline{\text{km of train-transport}} \\
\text{kg of carbon dioxide}
\end{pmatrix}
\tag{3.145}
$$

Then the consumption processes for driving a car and transport by train could be

$$
\mathbf{p_1} =
\begin{pmatrix}
-5 \\
-10^{-9} \\
10^{-9} \\
100 \\
0 \\
0 \\
0 \\
\underline{0} \\
10
\end{pmatrix}
\quad \text{and} \quad
\mathbf{p_2} =
\begin{pmatrix}
0 \\
0 \\
0 \\
0 \\
-100 \\
-10^{-12} \\
10^{-12} \\
\underline{10} \\
0
\end{pmatrix}
\tag{3.146}
$$

The vectors of final demand for car-driving and train-transport would then be

$$\mathbf{f}_1 = \begin{pmatrix} 0 \\ 0 \\ 0 \\ 1000 \\ 0 \\ 0 \\ 0 \\ 0 \end{pmatrix} \text{ and } \mathbf{f}_2 = \begin{pmatrix} 0 \\ 0 \\ 0 \\ 0 \\ 0 \\ 0 \\ 0 \\ 1000 \end{pmatrix} \tag{3.147}$$

Ignoring further upstream and downstream processes, the required scaling factors would be

$$\mathbf{s}_1 = \begin{pmatrix} 10 \\ 0 \end{pmatrix} \text{ and } \mathbf{s}_2 = \begin{pmatrix} 0 \\ 100 \end{pmatrix} \tag{3.148}$$

Section 3.8.1 discusses the computational aspects of product comparisons at length. For here is suffices to emphasise the importance of introducing a service-delivering flow, coming from the respective consumption processes, representing the reference flows for the respective product alternatives, and embodying two particular modes of supplying the functional unit to an exogenous consumer.

The concept of a reference flow is confusing in literature. Most discussions focus on performance characteristics, related to the way a consumer uses a product, like the dosage of washing powder that is used for washing a standard amount of clothes. We take the position presented by Guinée *et al.* (2002) that such information is part of the process data, and is therefore to be treated as an inventory concept. The reference flow is then the nature and amount of the service-delivering flow, emanating from a consumption process, which is assumed to be one mode of 'incarnation' of the functional unit. Thus, 1000 hour light might be a functional unit, while 1000 hour incandescent lamp-light is a possible reference flow.

3.7.3 Transport processes

Transport is an economic activity, with has inputs of fuels and capital equipment, and which is aimed at the shipping of a good from one location to another location. Its representation in flow charts for use in LCA is often unclear. As transport is between almost any two unit processes, it is often left out altogether, or one box is used to indicate transport, with arrows going to many of the other unit processes. It is therefore of interest to study the incorporation in the matrix structure in some detail.

One obvious way of including the aspect of transportation is to define an economic flow for the service km (or kg×km) of transport. Then there are aeroplanes, cars, trucks or trains which produce this service, and there are unit processes which require it as an input. For instance, the unit process traveling to a holiday resort requires a certain amount of this flow. However, more complicated is the situation that a unit processes, say production of tables, requires plastic, and that this plastic has to be transported to the appropriate location. It would be a bit strange to define the process of table production in terms of requiring a certain amount of plastic and a certain amount of transport of that plastic. A different representation starts from the idea that plastic at location X is different from plastic at location Y. The production of tables requires plastic at location X, and the production of plastic produces plastic at location Y. Then a transport process is a unit process which has plastic at location Y as an input and plastic at location X as an output. Other inputs are fuel and new trucks, other outputs are used trucks and carbon dioxide.

In practice, it may be convenient to use these two ways of representing transport processes in one system. Transport of a material from one factory to another could then employ the second type of representation, while transport of persons by train or car could employ the first type of representation.

3.7.4   Storage processes

While transport is an activity that converts a product from location X into a product at location Y, storage is an activity that converts a product at time X into a product at time Y. The same principles as in Section 3.7.3 can therefore be applied to storage processes. Here the input typically consists of cooling, or other ways of conservation.

Due to degradation (*e.g.*, by corrosion) or damage (*e.g.*, by insects) the amount present at time Y may be less than the amount at time X. Parts of the good may also disappear during storage and show up as an atmospheric emission of the storage process. Occasionally, quality differences (for the better or the worse) may occur during storage that force one to assign different names (and hence to assign different rows) to the good at time X and at time Y. This happens for instance in storage of cheese.

3.7.5   Waste treatment processes

Several times, in Sections 3.1 and 3.7.1, economic flows with the connotation of waste were encountered. Used generators and used cars are material products that are undesired. No one will buy a used generator or a used car in the meaning of one that is broken, and that should be treated in accordance with the legislation on waste. Such products have in general a negative economic value: the owner has to pay to carry it over to a waste treatment process.

It is then clear that the system should in principle contain a waste treatment process, that has the used generator or used car as an input. Possible other inputs are fuel, electricity, outputs comprise atmospheric pollutants and sometimes recycled material or recovered energy. See also Section 3.2 for a general discussion of possible multifunctionality problems involved in such unit processes. At this place, it is important to emphasise the structure of material products with a negative value, that flow from a production or consumption process towards a waste treatment process.

An alternative formulation would abandon the concept of material products with a negative value, and replace it by the service waste processing. A consumption process would in that case not have a used car as an output, but require the input of a service: used car demolition. The waste treatment process would not have used cars as an input, but the service of used car demolition as an output. When consistently followed, this structure would produce identical results as the former one. It is perhaps more likely to lead to errors in system construction. The formulation in terms of material products is expected to lead to a clearer picture of the flows in the system.

We admit, however, that the material point of view of waste processing will not in all cases lead to a sensible representation. For instance, the process of office cleaning is perhaps better conceived as providing the service of cleaning than as absorbing a certain amount of dust and dirt.

3.7.6   Waste flows

As several of the special processes discussed above make clear, waste flows can be seen as running between production or consumption processes and waste treatment processes. In following this interpretation, one adheres to the view that the flow of processes reflects a certain temporal ordering, where waste treatment is something that happens 'downstream', *i.e.* after the waste has been produced.

It is, however, possible to develop a different interpretation and representation for waste flows; see the previous section. That interpretation is based upon the idea that waste treatment facilities provide a function: getting rid of waste. The example on car-driving in Section 3.7.1 would then be reformulated in a linear space with basis

$$\begin{pmatrix} \text{litre of fuel} \\ \text{new cars} \\ \text{demolition of used cars} \\ \text{km of car-driving} \\ \overline{\text{kg of carbon dioxide}} \end{pmatrix} \tag{3.149}$$

so that the unit process that represents driving 100 km by car would look like

$$\mathbf{p} = \begin{pmatrix} -5 \\ -10^{-9} \\ -10^{-9} \\ 100 \\ \overline{10} \end{pmatrix} \tag{3.150}$$

It has not an output of 10^{-9} used car, but an input of 10^{-9} units of the service of demolition of used cars.

When the definition of the waste treatment is changed in consistency with this change of basis, the system looks similar, with the difference that all numbers in the row representing demolition of used cars have changed sign. The inventory vector remains – of course – unaffected.

The choice between representing waste as a flow or the service of waste treatment as a flow is one of psychological arguments. A number of such arguments are as follows, all relating to the case of representing the service of waste treatment instead of the flow of waste:

- it is no longer necessary to partition economic flows into goods and wastes;

- the temporal ordering of the processes flow diagram is distorted;

- the multifunctionality of a process is immediately visible in the form of more than output;

- the mass balance of a process is more difficult to check.

In this book, we have chosen for representing the physical flow, *i.e.* waste, instead of representing the service, *i.e.* treatment of waste. This is not fundamental, and one could easily reformulate the text in accordance with the second interpretation.

3.7.7   Environmental flows

Economic flows were discussed to be distinguishable into goods and wastes. Moreover, qualifications for temporal and regional differentiation of flows and brands of flows were discussed. For environmental flows the situation is similar but not exactly identical.

We can still distinguish environmental flows according to the region or the time period in which they are emitted or extracted. We can not distinguish goods and wastes or brands. But there are other aspects of interest here. It is convenient to distinguish several emission compartments, like air, water and soil, so that, for instance, mercury to air and mercury to water receive different rows in the intervention matrix. In impact assessment (see Section 8.1), they will often have different characterisation factors, so that a separation in the inventory analysis is needed.

Another distinction that may be useful is that between extractions of resources (often identified with inputs) and emissions of chemicals (often identified with outputs). Because emissions may occasionally show up as inputs of a process (and then be interpreted as a negative emission, like in sequestration of carbon dioxide by forestry) and resources as outputs (like in reintroduction programmes of wild animals), it may occasionally be useful to distinguish resources from emissions in a more explicit way than by the mere sign in the intervention matrix.

Finally, the distinction of substance speciation is one that may induce one to define a finer subdivision of environmental flows. For instance, Cr^{3+} and Cr^{6+} are two different forms of chromium with important differences in toxicological effects. Likewise, different forms of mercury may be of interest as well. The degree of refinement will be the result of an interplay between the availability of process data and the possibility for a separate treatment in impact assessment.

3.8   More than one reference flow

So far, we have discussed the case of analysing one product system with a definite final demand vector. It happens frequently, however, that one wishes to analyse more than one product system, or that an analysis of two or more final demand vectors, corresponding to two or more reference flows, is required. Two such instances occurred already in Sections 3.4.2 and 3.7.2. This section discusses two situations in which this occurs.

3.8.1 The comparison of product alternatives

Instead of defining one single final demand vector $\mathbf{f}$, several of such vectors $\mathbf{f}_1, \mathbf{f}_2, \ldots$ may be selected. The inventory problem

$$\mathbf{As} = \mathbf{f} \qquad (3.151)$$

then becomes a series of inventory problems:

$$\begin{aligned} \mathbf{As}_1 &= \mathbf{f}_1 \\ \mathbf{As}_2 &= \mathbf{f}_2 \\ &\cdots \end{aligned} \qquad (3.152)$$

Such a series of systems of equations may conveniently be written as

$$\mathbf{AS} = \mathbf{F} \qquad (3.153)$$

where

$$\mathbf{F} = \left(\ \mathbf{f}_1 \ \middle|\ \mathbf{f}_2 \ \middle|\ \cdots\ \right) \qquad (3.154)$$

and

$$\mathbf{S} = \left(\ \mathbf{s}_1 \ \middle|\ \mathbf{s}_2 \ \middle|\ \cdots\ \right) \qquad (3.155)$$

While the solution for a square invertible technology matrix $\mathbf{A}$ is given by

$$\mathbf{s} = \mathbf{A}^{-1}\mathbf{f} \qquad (3.156)$$

the solution to such a series of matrix equations may be expressed as

$$\mathbf{S} = \mathbf{A}^{-1}\mathbf{F} \qquad (3.157)$$

A similar augmentation of the inventory vector may be applied by replacing

$$\mathbf{g} = \mathbf{Bs} \qquad (3.158)$$

by

$$\mathbf{G} = \mathbf{BS} \qquad (3.159)$$

where

$$\mathbf{G} = \left(\ \mathbf{g}_1 \ \middle|\ \mathbf{g}_2 \ \middle|\ \cdots\ \right) \qquad (3.160)$$

Notice that this implies

$$\mathbf{G} = \mathbf{BA}^{-1}\mathbf{F} = \mathbf{\Lambda F} \qquad (3.161)$$

so that the intensity easily links a number of final demand vectors with a number of inventory vectors. Finally, the system vector $\mathbf{q}_1, \mathbf{q}_2, \ldots$ may be arranged into $\mathbf{Q}$:

$$\mathbf{Q} = \left(\begin{array}{c|c|c} \mathbf{q}_1 & \mathbf{q}_2 & \cdots \end{array} \right) = \left(\frac{\mathbf{F}}{\mathbf{G}} \right) \qquad (3.162)$$

but this is a form that will be of little use.

It is clear that the inventory problem, formulated in terms of $\mathbf{As}_k = \mathbf{f}_k$ and solved as $\mathbf{f}_k = \mathbf{A}^{-1}\mathbf{s}_k$ has one and the same technology matrix for every product alternative k. Moreover, it involves one single matrix inverse $\mathbf{A}^{-1}$. It is also clear that the expression $\mathbf{g}_k = \mathbf{\Lambda f}_k$ involves one unique matrix product: $\mathbf{\Lambda} = \mathbf{BA}^{-1}$, where $\mathbf{A}$ and $\mathbf{B}$ and hence the intensity matrix $\mathbf{\Lambda}$ have no index for the subscript k. This implies that several product alternatives may be analysed simultaneously, without the need to redo the construction of the technology matrix, the decisions and manipulations as to cut-off and allocation, and so on. Elements of the matrix $\mathbf{G}$ may be compared across columns, to compare product alternatives. For instance, the quantity $g_{11} - g_{12}$ measures the difference between system 1 and system 2 for the first environmental flow. One would be tempted to consider a positive difference as a beneficial judgement for system 2. However, this also depends on the meaning and environmental significance of the flow. For instance, such a positive value could indicate an advantage of system 1 for natural resources. The simultaneous analysis of several alternative systems is especially interesting in the context of a statistical analysis. See Section 6.4 for a discussion.

3.8.2   A database of inventory tables

A special case occurs when one wishes to make available the aggregated inventory tables for many different types of reference flow. For example, when the first flow represents litre of fuel and the second flow represents kWh of electricity, the final demand vector $\mathbf{f}_1 = \left(\begin{array}{cc} 1 & 0 \end{array} \right)^{\mathrm{T}}$ represents the reference flow 1 litre of fuel and the final demand vector $\mathbf{f}_2 = \left(\begin{array}{cc} 0 & 1 \end{array} \right)^{\mathrm{T}}$ represents the reference flow 1 kWh of electricity. The corresponding vectors $\mathbf{g}_1 = \mathbf{\Lambda f}_1$ and $\mathbf{g}_2 = \mathbf{\Lambda f}_2$ represent their inventory vectors. If we merge the two final demand vectors into

$$\mathbf{F} = \left(\begin{array}{c|c|c} \mathbf{f}_1 & \mathbf{f}_2 & \cdots \end{array} \right) = \mathbf{I} \qquad (3.163)$$

the resulting matrix

$$\mathbf{G} = \left(\begin{array}{c|c|c} \mathbf{g}_1 & \mathbf{g}_2 & \cdots \end{array} \right) \qquad (3.164)$$

represents the associated inventory vectors. $\mathbf{G}$ can thus be interpreted as a database of inventory tables. Because $\mathbf{F}$ is equal to the identity matrix $\mathbf{I}$, and in general $\mathbf{G} = \boldsymbol{\Lambda}\mathbf{F}$, $\mathbf{G}$ can simply be written as the intensity matrix:

$$\mathbf{G} = \boldsymbol{\Lambda} \qquad (3.165)$$

For obvious reasons, we will write $\mathbf{F_I}$ and $\mathbf{G_I}$ for these two special matrices:

$$\mathbf{F_I} = \mathbf{I} \qquad (3.166)$$

and

$$\mathbf{G_I} = \boldsymbol{\Lambda} \qquad (3.167)$$

The interesting feature is that an arbitrary final demand vector $\mathbf{f}$ can be written as a linear combination of the column vectors of $\mathbf{F_I}$. For instance, the final demand vector

$$\begin{pmatrix} 10 \\ 5 \end{pmatrix} \qquad (3.168)$$

can be written as

$$\begin{pmatrix} 1 & 0 \\ 0 & 1 \end{pmatrix} \begin{pmatrix} 10 \\ 5 \end{pmatrix} \qquad (3.169)$$

Obviously,

$$\mathbf{f} = \mathbf{F_I}\mathbf{f} \qquad (3.170)$$

Given the matrix $\mathbf{G_I}$, the inventory table associated with this choice of $\mathbf{f}$ is simply

$$\mathbf{g} = \mathbf{G_I}\mathbf{F_I}\mathbf{f} = \mathbf{G_I}\mathbf{f} = \boldsymbol{\Lambda}\mathbf{f} \qquad (3.171)$$

This shows that the matrix $\mathbf{G_I}$ or $\boldsymbol{\Lambda}$ can be used for a straightforward computation of the inventory table $\mathbf{g}$ of an arbitrary final demand vector $\mathbf{f}$. This idea has been employed by Frischknecht *et al.* (1993). A further interpretation of the intensity matrix $\boldsymbol{\Lambda}$ has been placed in Section 2.7.

3.8.3   Structural universality and sleeping processes

The fact that different product systems are computed with the same technology matrix and intervention matrix implies that the structure of the economy-environment is the same for every product alternative. Hence, the (qualitative) flow diagram of processes is equal for every product alternative considered. The only difference between product alternatives is that a different reference flow is used in the final demand vector, leading to a different scaling factors. Hence, one might say that all process flow diagrams are qualitatively the same, and that they only differ in quantitative terms. The structural equivalence of alternative product systems leads to an interesting analysis of this structure, as discussed in Chapter 7.

There is another item of interest in connection with this structural universality. It may happen that certain processes do not play a role in a certain LCA. For instance, it may well be the case that a reference flow for electricity does not make an appeal to the production of wheat. The scaling factor for the process of production of wheat will then be zero, and there is no need to give such a sleeping process a special treatment. It is not necessary to exclude this process from the system. For instance, when we add to the old example a third flow, representing kg of wheat, and a third process, production of wheat, which may or may not use the existing flows, we find a new technology matrix:

$$\mathbf{A} = \begin{pmatrix} -2 & 100 & 0 \\ 10 & 0 & -1 \\ 0 & 0 & 100 \end{pmatrix} \tag{3.172}$$

It may be inverted, and the scaling factor for the third process when requiring a final demand of $\mathbf{f} = \begin{pmatrix} 0 & 1000 & 0 \end{pmatrix}^{\mathrm{T}}$ is indeed $s_3 = 0$.

In fact, sleeping processes will occur quite often in product comparisons. When one compares an incandescent lamp with a fluorescent lamp, the first product system will have a sleeping process of fluorescent lamp production, while the second one will have a sleeping process of incandescent lamp production. As shown above, such cases need not bother us, and there is no reason for defining two technology matrices, one without fluorescent lamp production and one without incandescent lamp production. One and the same technology matrix may be used, and it is in fact advisable to do so, in relation with advanced types of analyses (like the discernibility analysis of Section 8.2.7 and the structural analysis of Chapter 7).

The automatic cut-off procedure, described in Section 3.3.2, also works fine. In the example, the third flow occurs with a positive coefficient, never with a negative coefficient. Because wheat is a good and not a waste, this does not meet the criteria for automatic cut-off stated in Section 3.3.4.

There are situations, however, in which sleeping processes might be of concern. This is especially the case when the sleeping process is multifunctional. When this happens, one would enter an allocation procedure, while a scaling factor of 0 would obviate the need for allocation. One solution to this is the removal of such processes from the technology matrix. Another solution involves the use of the pseudoinverse; see Section 3.2.5. Finally, it should be mentioned that the choice of the allocation principle for sleeping processes does not affect the results obtained. Although allocation can be avoided here, it does not harm as well.

3.9 Types of final demand

In Section 2.2, it was noted that many different choices for the final demand vector **f** are possible. From a computational point of view, any choice of **f** is fine. There are, however, different interpretations belonging to different forms of **f**. These differences relate to the values of the elements of **f** in relation to the meaning of the basis of the linear space in which is situated.

In most practical situation, there will only be one non-zero element in **f**. We have identified this element, say f_r, as the system's reference flow, ϕ. We will momentarily restrict the discussion to the case that all other elements of **f** indeed are zero. This restriction will be released in Section 3.9.4.

3.9.1 Cradle-to-grave analysis

We start by reconsidering the example of Section 3.7.1 and assume that the reference flow is 100 km of car-driving. This is a non-material service, that is produced by a system and that can be 'enjoyed' by a consumer. It, amongst others, involves the use of a car, which implies the need to produce a new car, as well as the need for waste treatment of a used car, even though this is only a tiny fraction of a car. The point is that no material product leaves the system (except of possible economic flows due to cut-off). A system in which only a non-material service leaves the system as a reference flow is normally interpreted as a cradle-to-grave analysis. All material flows are produced and treated inside the system. This is the most usual type of LCA, and it is the type of LCA that truly can be seen as covering an entire life cycle.

Other examples of reference flows of a cradle-to-grave analysis include electricity, having cut your hair, having had a bath, having had a good dinner and having seen a movie on TV.

3.9.2 Cradle-to-gate analysis

It can also be that the reference flow is an output of a material product with a positive economic value, *i.e.* a good, such as a TV, an electromotor, steel, PVC or a banana. In those cases, we do not investigate the grave of the product. The usual interpretation is that the product is followed until the gate of the 'factory' where it is produced or sold or, sometimes, until the household where it is consumed. The upstream processes are in principle included. The term cradle-to-gate analysis is often used to indicate such LCAs. Strictly speaking, such analyses do not cover the entire life cycle, and are therefore not an LCA *pur sang*. Notice that the reference flow is not a service-delivering flow, as in Section 3.7.2, but a material object.

3.9.3 Gate-to-grave analysis

The opposite of a cradle-to-gate analysis is also possible. The reference flow in such a case is the input of a material product with a negative value, a 'bad.' Examples include chemical waste, household waste and waste water. The system investigated is in those cases a waste-treatment system, and may include dumpsites, incinerators, waste water treatment plants, and so on. An appropriate name for such an analysis could be a gate-to-grave analysis. This name is normally not used. The idea of an LCA for waste treatment systems is found at many places in literature, however. Loosely speaking, one can say that a cradle-to-grave analysis and a gate-to-grave analysis add up to a cradle-to-gate analysis, although one might then miss (or duplicate perhaps) the consumption process. Again, the reference flow is a material object, not a service.

3.9.4 More general analyses

Above, three special cases have been distinguished: output of a non-material service, output of a material product with a positive value, and input of a material product with a negative value. We did not discuss the input of a non-material service, the input of a material product with a positive value, or the output of a material product with a negative value. Neither did we discuss the mixed case of several inputs and/or outputs of possibly different natures. It is difficult to assign sensible meanings to such final demand vectors in a general way. Occasionally, it may be needed to investigate such systems, most notably when the subject of interest is a bundle of commodities, as is for instance the case in studying the environmental consequences of entire households, societies or countries. We then start to use LCA for the analysis of scenarios. The mathematical formulation in terms of an arbitrary final demand vector then provides the necessary apparatus, although it may be questioned if the basic assumptions of the LCA-model as used here, which lead to the use of linear steady-state models, are applicable to such large-scale scenario studies.

3.10 General formulation of the refined model for inventory analysis

The previous sections have demonstrated that the basic model of Chapter 2 cannot be directly applied in certain cases. Especially, problems arise in connection to cut-off (Section 3.1), multifunctionality and allocation (Section 3.2), and closed-loop recycling (Section 3.5). This section concludes this chapter with a general formalism that includes these special situations.

Axiom 3 *Economic flows can be divided into two sets: goods, which have a positive utility, and wastes, which have a negative or zero utility.*

This definition is needed to be able to define and deal with multifunctional processes. One might want to replace the word utility by price or value. We have used the somewhat vaguer term utility because it leaves some freedom to specify what price or value is meant: market prices, shadow prices, prices corrected for taxes or subsidies, etc. In most practical cases, the division will be easy: steel, electricity, newspapers and cleaning services are goods and household waste, sewage effluent and used batteries are waste.

Definition 6 *An economic flow is said to be cut-off when, in case of a good, all coefficients for that flow in the technology matrix are non-positive, and in case of a waste, all coefficients for that flow in the technology matrix are non-negative.*

The rationale of this definition is as follows. A good i that is absorbed by one or more processes but not produced by a process has only coefficients a_{ij} that are zero or negative. This applies, for instance, to capital goods that are needed but for which no production data are available. The opposite case is represented by waste flows that are generated but for which no treatment processes have been described. There, the coefficients a_{ij} are zero or positive.

Definition 7 *A process is said to be multifunctional when it absorbs two or more wastes, produces two or more goods, or absorbs one or more wastes and produces one or more goods. Otherwise, it is said to be monofunctional.*

The discussion of Sections 3.4.3 and 3.8 show that we run into problems when there one good is produced by more than one process. In this case, of which we might say that there are several brands of that particular good, we would find several scaling factors that may be adjusted to produce a certain demand, leading to a problem of indeterminacy. It would be nice if we could derive a theorem on this subject, but we have not been able to do so, and therefore are forced to postulate the following axiom.

Axiom 4 *A technology matrix should not contain two or more processes that produce the same good or that absorb the same waste. This is equivalent to saying that each economic flow should come in one brand only.*

Lemma 2 *A technology matrix is square if and only if no economic flows have been cut-off and no process is multifunctional. It has more rows than columns if and only if economic flows have been cut-off and/or multifunctional process have been included.*

Proof   From Axiom 4, Definition 6 and Definition 7 it follows that in absence of cut-off and multifunctionality, every process produces one good or absorbs one waste, every good is produced by one process and every waste is absorbed by one process. This corresponds to a square technology matrix. Conversely, in a square technology matrix, there is exactly one process to produce each good or to absorb each waste. There are thus no flows left to account for cut-off or multifunctionality. This proofs the first part. In case of cut-off, there are flows (rows) for which there is no process (column), and in case of multifunctionality, there are flows (rows) that have no monofunctional (column), but that share a process with another flow. This proofs the second part of the lemma. Q.E.D.

Conjecture 1 *A system with a technology matrix which has more rows than columns can be solved exactly in either of two ways: with the pseudoinverse of the technology matrix*

$$\mathbf{s} = \mathbf{A}^{+}\mathbf{f} \tag{3.173}$$

or with

$$\mathbf{s} = \mathbf{A}'^{-1}\mathbf{f}' \tag{3.174}$$

where $\mathbf{A}'$ is a square and invertible matrix that is derived from the technology matrix $\mathbf{A}$ and $\mathbf{f}'$ is derived from $\mathbf{f}$ by means of eliminating rows of $\mathbf{A}$ and $\mathbf{f}$ that correspond to cut-off, by adding columns to $\mathbf{A}$ to account for so-called avoided processes, and by replacing columns of $\mathbf{A}$ that represent multifunctional processes by a number of monofunctional processes.

Sketch of a proof First, we will enter $\mathbf{s} = \mathbf{A}^+\mathbf{f}$ into the equation $\tilde{\mathbf{f}} = \mathbf{A}\mathbf{s}$ (see also Lemma 1). This yields

$$\tilde{\mathbf{f}} = \mathbf{A}\mathbf{A}^+\mathbf{f} = \mathbf{A}\left(\mathbf{A}^{\mathrm{T}}\mathbf{A}\right)^{-1}\mathbf{A}^{\mathrm{T}}\mathbf{f} \tag{3.175}$$

Whence it follows that

$$\mathbf{A}^{\mathrm{T}}\mathbf{f} = \mathbf{A}^{\mathrm{T}}\tilde{\mathbf{f}} \tag{3.176}$$

From elementary linear algebra, it follows that $\mathbf{s} = \mathbf{A}^+\mathbf{f}$ may indeed occasionally be an exact solution to $\mathbf{A}\mathbf{s} = \mathbf{f}$. For the second part of the proposition, we do not have a formal proof yet, but have to rely on the ideas that were presented in Sections 3.1 and 3.2.

Clearly, this last conjecture should become a theorem in due time.

3.11   An extended example

This chapter has shown how the basic model is to be applied, interpreted and adapted for situations arising in real life. We conclude this chapter by showing how the various adaptations are to be applied in an example case. This example case is more complex than that discussed before. It contains two multifunctional processes, one treated with the substitution and one with partitioning method, it comprises two alternative product systems with two separate reference flows emanating from two consumption processes, it contains economic flows that are categorised as goods and wastes, and it contains flows to be cut-off. On the other hand, the example is not yet so complicated as a real example. A real example may include matrices with several hundreds of rows and columns. These are difficult to display, and perhaps even more difficult to interpret in words. Therefore this example presents a moderate-size hypothetical case. It is a modification of the example in Heijungs & Kleijn (2001). All calculations have been made with the CMLCA software.

The basis of the economic part of the linear space is

$$\begin{pmatrix} \text{incandescent lamps} \\ \text{MJ of electricity} \\ \text{disposed incandescent lamps} \\ \text{hr of incandescent lamp light} \\ \text{kg of glass} \\ \text{kg of copper} \\ \text{kg of fuel} \\ \text{fluorescent lamps} \\ \text{disposed fluorescent lamps} \\ \text{hr of fluorescent lamp light} \\ \text{MJ of heat} \\ \text{kg of recycled copper} \\ \text{kg of waste residue} \end{pmatrix} \qquad (3.177)$$

The partitioning of economic flows into goods and wastes is done such that the third and the ninth flow – disposed lamps – are wastes, as well as the 13th, waste residue. Thus, G={incandescent lamps, MJ of electricity, hr of incandescent lamp light, kg of glass, kg of copper, kg of fuel, fluorescent lamps, hr of fluorescent lamp light, MJ of heat, kg of recycled copper} and W={disposed incandescent lamps, disposed fluorescent lamps}.

The names of the processes is 'use of incandescent lamps', 'production of incandescent lamps', 'production of electricity', 'incineration of disposed incandescent lamps', 'production of glass', 'production of copper', 'production of fuel', 'use of fluorescent lamps', 'production of fluorescent lamps' and 'incineration of disposed fluorescent lamps'.

The technology matrix $\mathbf{A}$ is

$$\begin{pmatrix}
-1 & 1000 & 0 & 0 & 0 & 0 & 0 & 0 & 0 & 0 \\
-10000 & -1000 & 1 \times 10^6 & 0 & -100 & -10000 & 0 & -5000 & -3000 & 0 \\
1 & 0 & 0 & -100 & 0 & 0 & 0 & 0 & 0 & 0 \\
5000 & 0 & 0 & 0 & 0 & 0 & 0 & 0 & 0 & 0 \\
0 & -10 & 0 & 0 & 1000 & 0 & 0 & 0 & -20 & 0 \\
0 & -5 & 0 & 0 & 0 & 100 & 0 & 0 & -150 & 0 \\
0 & 0 & -500 & 0 & 0 & 0 & 1000 & 0 & 0 & 0 \\
0 & 0 & 0 & 0 & 0 & 0 & 0 & -1 & 1000 & 0 \\
0 & 0 & 0 & 0 & 0 & 0 & 0 & 1 & 0 & -100 \\
0 & 0 & 0 & 0 & 0 & 0 & 0 & 25000 & 0 & 0 \\
0 & 0 & 2 \times 10^6 & 0 & 0 & 0 & 0 & 0 & 0 & 0 \\
0 & 0 & 0 & 0.5 & 0 & 0 & 0 & 0 & 0 & 0 \\
0 & 0 & 0 & 0 & 0 & 0 & 0 & 0 & 0 & 2
\end{pmatrix}$$

$$(3.178)$$

From these definitions and coefficients, it appears that flow 13 (waste residue) is a waste that is produced but not treated, thus must be subject to cut-off. Furthermore, there are two multifunctional processes: the third (production of electricity) and the fourth (incineration of disposed incandescent lamps). The first of these will be treated with the partitioning method, allocating 0.8 to electricity and 0.2 to heat. The second one will be treated with the substitution method, considering recycled copper as equivalent to copper, with a correction factor of 0.9 to account for differences in quality. Although process 1, for instance has 2 outputs, it is not a multifunctional process, because only one of these outputs is a good. The technology matrix after cut-off and allocation $\mathbf{A}'$ becomes:

$$
\begin{pmatrix}
-1 & 1000 & 0 & 0 & 0 & 0 & 0 & 0 & 0 & 0 & 0 \\
-10000 & -1000 & 1 \times 10^6 & 0 & 0 & -100 & -10000 & 0 & -5000 & -3000 & 0 \\
1 & 0 & 0 & 0 & -100 & 0 & 0 & 0 & 0 & 0 & 0 \\
5000 & 0 & 0 & 0 & 0 & 0 & 0 & 0 & 0 & 0 & 0 \\
0 & -10 & 0 & 0 & 0 & 1000 & 0 & 0 & 0 & -20 & 0 \\
0 & -5 & 0 & 0 & 0.45 & 0 & 100 & 0 & 0 & -150 & 0 \\
0 & 0 & -400 & -100 & 0 & 0 & 0 & 1000 & 0 & 0 & 0 \\
0 & 0 & 0 & 0 & 0 & 0 & 0 & 0 & -1 & 1000 & 0 \\
0 & 0 & 0 & 0 & 0 & 0 & 0 & 0 & 1 & 0 & -100 \\
0 & 0 & 0 & 0 & 0 & 0 & 0 & 0 & 25000 & 0 & 0 \\
0 & 0 & 0 & 2 \times 10^6 & 0 & 0 & 0 & 0 & 0 & 0 & 0
\end{pmatrix}
$$

$$(3.179)$$

This matrix is square and non-singular, hence it is invertible. The inverse $\mathbf{A}'^{-1}$ is occupies too much space to display here.

Let us specify the functional unit as '10 hr of light', and define two alternative reference flows: '10 hr of incandescent lamp light' and '10 hr of fluorescent lamp light'. These can be formalised as two alternative final demand vectors:

$$
\mathbf{f}_1 = \begin{pmatrix} 0 \\ 0 \\ 0 \\ 10 \\ 0 \\ 0 \\ 0 \\ 0 \\ 0 \\ 0 \\ 0 \\ 0 \end{pmatrix} \quad \text{and} \quad \mathbf{f}_2 = \begin{pmatrix} 0 \\ 0 \\ 0 \\ 0 \\ 0 \\ 0 \\ 0 \\ 0 \\ 10 \\ 0 \\ 0 \\ 0 \end{pmatrix}
\qquad (3.180)
$$

These final demand vectors are modified in accordance with the cut-off and allocation operations, and change into

$$\mathbf{f}'_1 = \begin{pmatrix} 0 \\ 0 \\ 0 \\ 10 \\ 0 \\ 0 \\ 0 \\ 0 \\ 0 \\ 0 \\ 0 \end{pmatrix} \quad \text{and} \quad \mathbf{f}'_2 = \begin{pmatrix} 0 \\ 0 \\ 0 \\ 0 \\ 0 \\ 0 \\ 0 \\ 0 \\ 10 \\ 0 \end{pmatrix} \tag{3.181}$$

Multiplication of the inverse of the modified technology matrix with the two respective final demand vectors yields the scaling vectors for both alternative systems:

$$\mathbf{s}'_1 = \begin{pmatrix} 0.002 \\ 2 \times 10^{-6} \\ 2 \times 10^{-5} \\ 0 \\ 2 \times 10^{-5} \\ 2 \times 10^{-8} \\ 1 \times 10^{-8} \\ 8 \times 10^{-6} \\ 0 \\ 0 \\ 0 \end{pmatrix} \quad \text{and} \quad \mathbf{s}'_2 = \begin{pmatrix} 0 \\ 0 \\ 2 \times 10^{-6} \\ 0 \\ 0 \\ 8 \times 10^{-9} \\ 6 \times 10^{-7} \\ 8 \times 10^{-7} \\ 0.0004 \\ 4 \times 10^{-7} \\ 4 \times 10^{-6} \end{pmatrix} \tag{3.182}$$

The basis of the environmental part of the linear space is

$$\begin{pmatrix} \text{kg of carbon dioxide to air} \\ \text{kg of sulphur dioxide to air} \\ \text{kg of copper to soil} \\ \text{kg of sand} \\ \text{kg of copper ore} \\ \text{kg of crude oil} \end{pmatrix} \tag{3.183}$$

Note the indication of compartments 'to air' and 'to soil'. The intervention matrix $\mathbf{B}$ is in the unallocated form

$$
\begin{pmatrix}
0 & 0 & 1000 & 100 & 0 & 0 & 200 & 0 & 0 & 200 \\
0 & 0 & 100 & 0 & 0 & 0 & 5 & 0 & 0 & 0 \\
0 & 0 & 0 & 0.75 & 0 & 0 & 0 & 0 & 0 & 4 \\
0 & 0 & 0 & 0 & -1000 & 0 & 0 & 0 & 0 & 0 \\
0 & 0 & 0 & 0 & 0 & -1000 & 0 & 0 & 0 & 0 \\
0 & 0 & 0 & 0 & 0 & 0 & -1200 & 0 & 0 & 0
\end{pmatrix}
\tag{3.184}
$$

and after allocation, it is $\mathbf{B}'$:

$$
\begin{pmatrix}
0 & 0 & 800 & 200 & 100 & 0 & 0 & 200 & 0 & 0 & 200 \\
0 & 0 & 80 & 20 & 0 & 0 & 0 & 5 & 0 & 0 & 0 \\
0 & 0 & 0 & 0 & 0.75 & 0 & 0 & 0 & 0 & 0 & 4 \\
0 & 0 & 0 & 0 & 0 & -1000 & 0 & 0 & 0 & 0 & 0 \\
0 & 0 & 0 & 0 & 0 & 0 & -1000 & 0 & 0 & 0 & 0 \\
0 & 0 & 0 & 0 & 0 & 0 & 0 & -1200 & 0 & 0 & 0
\end{pmatrix}
\tag{3.185}
$$

Observe that cut-off does not affect $\mathbf{B}$, but that allocation does.

Multiplication of this matrix with the two scaling vectors yields the two inventory vectors:

$$
\mathbf{g}_1 =
\begin{pmatrix}
0.020 \\
0.0016 \\
1.5 \times 10^{-5} \\
-2 \times 10^{-5} \\
-1 \times 10^{-5} \\
-0.0096
\end{pmatrix}
\text{ and } \mathbf{g}_2 =
\begin{pmatrix}
0.0026 \\
0.00016 \\
1.6 \times 10^{-5} \\
-8 \times 10^{-6} \\
-0.0006 \\
-0.00096
\end{pmatrix}
\tag{3.186}
$$

The first three rows represent outputs of the system, the last three inputs.

The cut-off procedure effectively partitions $\mathbf{A}$ into $\mathbf{A}'$ and $\mathbf{A}''$, where the latter one is given by

$$
\begin{pmatrix} 0 & 0 & 0 & 0 & 0 & 0 & 0 & 0 & 0 & 2 \end{pmatrix}
\tag{3.187}
$$

Multiplied with the scaling factor for s_{11}, we find

$$
\mathbf{f}'' = \begin{pmatrix} 8 \times 10^{-6} \end{pmatrix}
\tag{3.188}
$$

Stacking $\mathbf{f}'$ and $\mathbf{f}''$ yields the final supply vectors for the two systems

$$
\tilde{\mathbf{f}}_1 = \begin{pmatrix} -9.4 \times 10^{-11} \\ 1.6 \times 10^{-6} \\ -8.7 \times 10^{-11} \\ 10 \\ 2.2 \times 10^{-12} \\ -1.9 \times 10^{-12} \\ -2.4 \times 10^{-10} \\ 0 \\ 0 \\ 0 \\ 0 \\ 0 \\ 0 \end{pmatrix} \quad \text{and } \tilde{\mathbf{f}}_2 = \begin{pmatrix} 0 \\ -5.7 \times 10^{-8} \\ 0 \\ 0 \\ -3.2 \times 10^{-13} \\ -4.2 \times 10^{-12} \\ 3.4 \times 10^{-11} \\ 1.5 \times 10^{-11} \\ 3.6 \times 10^{-11} \\ 10 \\ 0 \\ 0 \\ 8 \times 10^{-6} \end{pmatrix} \tag{3.189}
$$

so that the discrepancy vectors are

$$
\mathbf{d}_1 = \begin{pmatrix} -9.4 \times 10^{-11} \\ 1.6 \times 10^{-6} \\ -8.7 \times 10^{-11} \\ 0 \\ 2.2 \times 10^{-12} \\ -1.9 \times 10^{-12} \\ -2.4 \times 10^{-10} \\ 0 \\ 0 \\ 0 \\ 0 \\ 0 \\ 0 \end{pmatrix} \quad \text{and } \mathbf{d}_2 = \begin{pmatrix} 0 \\ -5.7 \times 10^{-8} \\ 0 \\ 0 \\ -3.2 \times 10^{-13} \\ -4.2 \times 10^{-12} \\ 3.4 \times 10^{-11} \\ 1.5 \times 10^{-11} \\ 3.6 \times 10^{-11} \\ 0 \\ 0 \\ 0 \\ 8 \times 10^{-6} \end{pmatrix} \tag{3.190}
$$

Clearly, the cut-off is not relevant for sleeping processes, and as process 13 is only involved for the second product alternative – fluorescent lamps – the cut-off causes only a discrepancy for that alternative. The other items, mostly around 10^{-11}, may be interpreted as resulting from round-off; see Section 6.6 for more information.

The intensity matrix $\mathbf{\Lambda'}$ follows from multiplication of $\mathbf{B'}$ and $\mathbf{A'}^{-1}$:

$$\mathbf{\Lambda'} = \begin{pmatrix}
0.0013 & 0.00012 & 0 & -0.01 & -0.05 & -0.00072 \\
0.00088 & 8.2 \times 10^{-5} & 0 & 0 & 0 & -0.00048 \\
-1 & 3.69 \times 10^{-5} & -0.0075 & 0 & -0.045 & -0.00021 \\
0.002 & 0.00016 & 1.5 \times 10^{-6} & -2 \times 10^{-6} & -1 \times 10^{-6} & -0.00096 \\
8.8 \times 10^{-5} & 8.2 \times 10^{-6} & 0 & -1 & 0 & -4.8 \times 10^{-5} \\
0.088 & 0.0082 & 0 & 0 & -10 & -0.048 \\
0.2 & 0.005 & 0 & 0 & 0 & -1.2 \\
0.016 & 0.0015 & 0 & -0.02 & -1.5 & -0.0086 \\
-2 & 0 & -0.04 & 0 & 0 & 0 \\
0.00026 & 1.6 \times 10^{-5} & 1.6 \times 10^{-6} & -8 \times 10^{-7} & -6 \times 10^{-5} & -9.6 \times 10^{-5} \\
0.00011 & 1 \times 10^{-5} & 0 & 0 & 0 & -6 \times 10^{-5}
\end{pmatrix}$$

$$(3.191)$$

We could, finally, pursue to construct and display the process matrix $\mathbf{P}$ and the system vector $\mathbf{q}$, but this does not add a clear value to this example.

The present example makes clear, amongst others, that a matrix formalism greatly facilitates the handling of large amounts of process data as occurs in a typical LCA. But it also demonstrates the problems involved in displaying and overviewing the large matrices that result from a matrix-oriented treatment. Chapter 7 will explore some preliminary techniques to overcome these problems.

Chapter 4

Advanced topics in inventory analysis*

This chapter deals with some quite sophisticated topics. The material is not essential for an understanding of the computational structure of LCA *per se*, but it points out some interesting connections to alternative approaches to LCA.

4.1 Alternative ways of formulating and solving the inventory problem

In Chapter 1, we observed that the computational structure of LCA is only seldom discussed in the standard books on LCA. It is therefore often impossible to describe the approach followed by other authors to formulate and solve the inventory problem. A similar argument applies even more strongly to software for LCA. Documentation of software for LCA is often geared towards using the program, and technical issues of implementation tend to be left out. Moreover, this documentation is often shipped with the software, at costs of several thousands of dollars or euros, so that these texts cannot be considered as ordinary public knowledge. Finally, in concrete LCA case studies, the computational approach employed is not discussed, or it is only referred to in terms of the software that was used.

In a limited number of cases, such information is available, to a greater or lesser extent. From this, it emerges that a formulation and solution in terms of matrices is not the only approach. It is even not the one used most frequently; see also Section 1.1.2. Some alternative approaches are

discussed in this section. It should be noted that the descriptions provided
are not always completely clear or complete.

Given the fact that the computational structure of inventory analysis
is not often discussed, it is not surprising that comparative analyses are
even harder to find. As far as known by the authors, Heijungs (1994) and
more recently Melo (1999) and Suh & Huppes (2002) are the only ones to
provide comparisons of a number of such approaches.

4.1.1   The sequential method

The most popular way of formulating and solving the inventory problem
seems to be the sequential method. Here, the scaling of the processes is
achieved not simultaneously but in a sequential way. Let us consider the
example of a reference flow of 1000 kWh of electricity. The first process
produces 10 kWh of electricity, hence we need to scale it by a factor of
100. This then leads to 100 kg of carbon dioxide, and to a demand for
200 litre of fuel. As the second process produces 100 litre of fuel, we need
to scale this process by a factor of 2. This gives another 20 kg of carbon
dioxide. For a large system, this procedure of moving further upstream
must be repeated many times. In addition, it must be carried out for
the downstream processes as well. After this the individual contributions
of the same environmental flow can be added. In the example, we find
$100 + 20 = 120$ kg of carbon dioxide.

The sequential method is trivial and easy to understand. It is however
not easy to formalise with mathematical notation. Some attempts have
been made; see, *e.g.*, Lübkert *et al.* (1991) and Anonymous (1995). The
formulation in terms of a matrix inversion is perhaps less trivial, but it has
the advantage of providing a formalism that can be subject to various sorts
of analysis; see for instance Chapter 6.

Another disadvantage of the sequential method is of a more practical
nature. When the linkage of the processes is not purely linear but is a
network which includes feedback loops, the sequential method becomes
difficult to apply. Feedback loops occur frequently in industrial systems.
For instance, mining of coal needs electricity, while production of electricity
needs coal. This is most often not regarded as a feedback loop, but more as
a mutual dependency. From a logical point of view, there is no difference
between feedback loops and mutual dependencies. The crucial element is
that the processes cannot be delineated in a linear way, but can instead
be said to bite in their own tail. Closed-loop recycling (see Section 3.5)
is another example of a situation in which feedback loops exists in the

system. Let us suppose that fuel production needs a certain amount of electricity. Then, to make electricity we need fuel and to make fuel we need electricity. To make this electricity, we again need fuel, and so on and so on. An infinite sequence of upstream processes would be involved. When the sequential method is implemented in this way, a computer program will not stop, except when the memory requirements exceed the available capacity, resulting in a message like "stack overflow" or "out of memory." To prevent such disastrous results, solutions have been proposed including:

- interrupting a branch after a specified number of loops (Fritsche *et al.* (1991));

- interrupting a branch when the last round has added less than a specified amount (Fava *et al.* (1991), Boustead (1993));

- replacing process data by 'corrected' process data, in which feedback loops have been accounted for (Häuslein & Hedemann (1995));

- use of infinite geometrical progression, in which an infinite sum is treated with methods from mathematical analysis (Suh & Huppes (2002)).

It will be clear that these are imperfect solutions, and that one must replace the idea of a linear network by that of a network of interlinked processes and move to a simultaneous solving of the equations in order to yield solutions that are more precise and do better justice to the fact that the economic reality indeed suggests a web of interlinked processes. It should be noted, however, that the solution of simultaneous equations in practical implementations is often based on iterative algorithms; see Section 10.2 for a brief discussion. Furthermore, Section 4.3.2 discusses how the sequential method with an infinite sum of ever-decreasing terms converges to the value that is obtained by matrix inversion.

4.1.2 Petri nets

A number of times, and especially in connection to the Umberto software, the use of Petri-nets is discussed as a way to approach the inventory problem (see, *e.g.*, Möller & Rolf (1995), Häuslein & Hedemann (1995), Schmidt & Haäuslein (1997) and Melo (1999)). The theory of Petri-nets provides a set of visual elements which can be used to define a system. The basic elements are places, which represent conditions or states, transitions, which represent events or activities, and arcs, which represent places with transitions or the other way around. From the few descriptions on the use of Petri-nets in relation to the inventory problem, we get the impression that the main advantage of Petri-nets is a strict set of rules for the visual representation of flow diagrams, but that the mathematical rules are close or perhaps even identical to the sequential case.

4.1.3 Linear programming

Linear programming has been mentioned in Section 3.2.5 as a possible way to address the problem of multifunctionality. Azapagic & Clift (1994, 1999) give it a wider use, namely in formulating and solving the inventory problem itself. Their basic equation may be written as

$$\mathbf{g} = \mathbf{Yf} \tag{4.1}$$

where $\mathbf{Y}$ is a matrix of so-called burden coefficients. This formulation then is embedded in a context of multi-objective optimisation, *e.g.*, to minimise costs and environmental impacts. Useful as this may be, the question remains how $\mathbf{Y}$ is measured or constructed. Recalling the results obtained so far, the only possible conclusion is that $\mathbf{Y}$ is identical to the intensity matrix $\boldsymbol{\Lambda}$. Indeed, the work by Azapagic & Clift (1999) can be used to incorporate life cycle assessment into a more general decision-theoretic framework, such as multiobjective decision-support. But it is clear that the specification of $\mathbf{Y}$ requires that a life cycle assessment be carried out prior to the inclusion in the decision-theoretic framework. For this, the present text offers explicit formulas when the connection $\mathbf{Y} = \boldsymbol{\Lambda}$ is made.

Somewhat more involved is the 'life cycle activity analysis' approach proposed by Freire *et al.* (2001), where the balance equations are integrated in a linear programming framework.

4.1.4 Cramer's rule

Heijungs (1992, 1994) formulates a matrix approach without the matrix inverse. It is based on Cramer's rule for solving a system of linear equations

$$\mathbf{As} = \mathbf{f} \tag{4.2}$$

with $\mathbf{A}$ square and non-singular. Cramer's rule provides a solution for s as

$$\forall k : s_k = \frac{\det(\mathbf{A}_k(\mathbf{f}))}{\det(\mathbf{A})} \tag{4.3}$$

where $\det(\cdot)$ is the determinant of a matrix, and $\mathbf{A}_k(\mathbf{f})$ is matrix $\mathbf{A}$ with the kth column replaced by $\mathbf{f}$. The advantage of this formulation was the feeling that a determinant is easier to compute than an inverse matrix, and that the formulation in terms of determinants can be used in deriving perturbation-theoretic concepts. In the last decade, the capacities of personal computers have increased to an extent that matrix inversion of large matrices presents no special problems. Furthermore, more research in this field has opened up the area of perturbation theory for inverse matrices; see Chapter 6. These developments have obviated the use of Cramer's rule.

4.2   Expansion of the inverse as a power series

With the rise of input-output analysis (for more information, see Section 5.1), the inversion of large matrices became an aspect of concern. Whereas the first input-output study was restricted to a 9×9-dimensional matrix, the availability of data soon leaded to matrices with one hundred or more rows and columns. As computers were at that time hardly available, it is natural that computational shortcuts were developed to circumvent the cumbersome operation of inverting such a matrix. One such shortcut is of more than algorithmical interest, as it provides an interpretation in terms of successive orders of dependency of the system of equations. Moreover, it has been discussed within the context of LCA as well (Frischknecht & Kolm, 1995; Schmidt, 1995).

It is well known that for any real- or complex-valued x with $|x| < 1$, one may write

$$\frac{1}{1-x} = 1 + x + x^2 + \cdots = \sum_{n=0}^{\infty} x^n \tag{4.4}$$

Under the condition specified,

$$\lim_{x \to \infty} x^n = 0 \tag{4.5}$$

This means that a power series of a finite number of terms may be used to approximate the quantity indicated.

There exists a powerful generalisation when the scalar x is replaced by a matrix $\mathbf{X}$, the reciprocal operation being replaced by an inversion:

$$(\mathbf{I} - \mathbf{X})^{-1} = \mathbf{I} + \mathbf{X} + \mathbf{X}^2 + \cdots = \sum_{n=0}^{\infty} \mathbf{X}^n \tag{4.6}$$

where $\mathbf{I}$ is the unit matrix of an appropriate dimension. Its proof will be reproduced from Chiang (1984, p. 120–121), because it is illustrative for the conditions under which this equation is true. This proof is based on the multiplication of $\mathbf{I} - \mathbf{X}$ and the finite series $\mathbf{I} + \mathbf{X} + \mathbf{X}^2 + \cdots + \mathbf{X}^n$:

$$
\begin{aligned}
&(\mathbf{I} - \mathbf{X})\left(\mathbf{I} + \mathbf{X} + \mathbf{X}^2 + \cdots + \mathbf{X}^n\right) = \\
&\left(\mathbf{I} + \mathbf{X} + \mathbf{X}^2 + \cdots + \mathbf{X}^n\right) - \mathbf{X}\left(\mathbf{I} + \mathbf{X} + \mathbf{X}^2 + \cdots + \mathbf{X}^n\right) = \\
&\left(\mathbf{I} + \mathbf{X} + \mathbf{X}^2 + \cdots + \mathbf{X}^n\right) - \left(\mathbf{X} + \mathbf{X}^2 + \mathbf{X}^3 + \cdots + \mathbf{X}^{n+1}\right) = \\
&\mathbf{I} - \mathbf{X}^{n+1}
\end{aligned}
\tag{4.7}
$$

Under the condition that $\mathbf{X}^{n+1}$ converges to $\mathbf{0}$ for sufficiently large n, this expression reduces to

$$
(\mathbf{I} - \mathbf{X})\left(\mathbf{I} + \mathbf{X} + \mathbf{X}^2 + \cdots + \mathbf{X}^n\right) = \mathbf{I}
\tag{4.8}
$$

This means that, for sufficiently large n

$$
(\mathbf{I} - \mathbf{X})^{-1} = \mathbf{I} + \mathbf{X} + \mathbf{X}^2 + \cdots + \mathbf{X}^n
\tag{4.9}
$$

A sufficient condition for convergence is that the elements of $\mathbf{X}$ are non-negative (*i.e.*, $\mathbf{X}$ must be positive semi-definite), and that the elements in each column add up to less than 1, which can alternatively be stated as that the norm of $\mathbf{X}$ be less than 1, or that the spectral radius of $\mathbf{X}$ be less than 1. See Waugh (1950), Atkinson (1989, p.491) or Miller & Blair (1995, p.22 *ff.*) for a number of more formal proofs. For a normal input-output transactions matrix, these requirements are fulfilled by the very nature of the input-output table.

For the technology matrix in LCA, however, things are different. Frischknecht & Kolm (1995) and Schmidt (1995) discuss the use of power series for approximating the inverse of the technology matrix. They replace $\mathbf{I} - \mathbf{X}$ by $\mathbf{A}$ so that

$$
\mathbf{A}^{-1} = \mathbf{I} + (\mathbf{I} - \mathbf{A}) + (\mathbf{I} - \mathbf{A})^2 + \cdots + (\mathbf{I} - \mathbf{A})^n
\tag{4.10}
$$

They are fortunate in having an example system for which the approximation converges indeed. Had they chosen the simple system that plays a central role in this book, they could not have applied the approximation. The technology matrix is

$$
\mathbf{A} = \begin{pmatrix} -2 & 100 \\ 10 & 0 \end{pmatrix}
\tag{4.11}
$$

so that one has

$$\mathbf{I} - \mathbf{A} = \begin{pmatrix} 3 & -100 \\ -10 & 1 \end{pmatrix} \qquad (4.12)$$

This yields the power expansion that begins with

$$\begin{pmatrix} 1 & 0 \\ 0 & 1 \end{pmatrix} + \begin{pmatrix} 3 & -100 \\ -10 & 1 \end{pmatrix} + \begin{pmatrix} 1009 & -400 \\ -40 & 1001 \end{pmatrix} + \\ \begin{pmatrix} 7027 & -101300 \\ -10130 & 5001 \end{pmatrix} + \cdots \qquad (4.13)$$

It clearly diverges instead of converging to

$$\mathbf{A} = \begin{pmatrix} 0 & 0.1 \\ 0.01 & 0.002 \end{pmatrix} \qquad (4.14)$$

We conclude that, although the expansion as a power series may work in certain cases, its use in LCA cannot be recommended as a general solution. Fortunately, the capacity of computers has increased so rapidly that matrix inversion nowadays is not likely to create computational problems in terms of time and memory constraints, provided that efficient routines are used (for this, see Section 10.2.1).

The applicability of the ideas presented in this section can be increased when rescaling procedures of $\mathbf{A}$ are employed. This is discussed in Section 4.3.4.

4.3 Feedback loops in the technology matrix

Let us take up the very first example of the inventory problem. It was characterised with a technology matrix

$$\mathbf{A} = \begin{pmatrix} -2 & 100 \\ 10 & 0 \end{pmatrix} \qquad (4.15)$$

and a final demand vector

$$\mathbf{f} = \begin{pmatrix} 0 \\ 1000 \end{pmatrix} \qquad (4.16)$$

The balance equation were also presented in the form

$$\begin{cases} -2 \times s_1 + 100 \times s_2 = 0 \\ 10 \times s_1 + 0 \times s_2 = 1000 \end{cases} \qquad (4.17)$$

In this particular case, there was in fact not so much need to employ the inverse of a matrix in solving the balance equations. The second equation gave a straightforward solution for s_1: 100. This could then be used to compute a solution for s_2: 2. We may interpret this as applying a sequential method; see Section 4.1.1.

The sequential method is intuitively easy. A solution is found by scaling the unit processes one by one. This, however, requires that the unit processes can be arranged in an order which permits this step-by-step solution. This is often not possible. In this section, we discuss more extensively the situation that feedback loops are present in the system.

Let us change the technology matrix into

$$\mathbf{A} = \begin{pmatrix} -2 & 100 \\ 10 & -10 \end{pmatrix} \tag{4.18}$$

That is, we assume that the production of electricity requires fuel and that the production of fuel requires electricity. In other words, there is a feedback loop in the system. We will now elaborate how the sequential method treats a square technology matrix that represents a system with feedback loops. In the next sections, results will be given in many digits in order to demonstrate how successive steps in an algorithm add more and more digits. For practical purposes, the number of digits shown suggests too much precision.

4.3.1   Solution with matrix inversion

The occurrence of feedback loops is a physical interpretation of an otherwise abstract mathematical description. The equation

$$\mathbf{As} = \mathbf{f} \tag{4.19}$$

remains valid, irrespective of whether $\mathbf{A}$ is interpreted as describing a system with feedback loops, and

$$\mathbf{s} = \mathbf{A}^{-1}\mathbf{f} \tag{4.20}$$

remains a valid solution, as long as $\mathbf{A}$ is square and invertible. Hence, the inverse matrix provides a straightforward solution. It is

$$\mathbf{A}^{-1} = \begin{pmatrix} 0.1020408\cdots & 0.1020408\cdots \\ 0.01020408\cdots & 0.0020408\cdots \end{pmatrix} \tag{4.21}$$

so that the scaling factors are easily found as

$$\mathbf{s} = \mathbf{A}^{-1}\mathbf{f} = \begin{pmatrix} 102.0408\cdots \\ 2.0408\cdots \end{pmatrix} \quad (4.22)$$

The fact that matrix-inversion-based solution of the inventory problem deals with feedback loops in an exact and easy way is one major motivation to develop matrix methods for LCA. It may be observed that many of the programs listed in Siegenthaler *et al.* (1997) have difficulties in dealing with systems containing feedback loops.

4.3.2 The network as an infinite sequence

With the technology matrix with a feedback loop, the system of equations is changed accordingly into

$$\begin{cases} -2 \times s_1 + 100 \times s_2 = 0 \\ 10 \times s_1 + -10 \times s_2 = 1000 \end{cases} \quad (4.23)$$

It is obvious that we cannot obtain a solution for s_1 as before. We can start to propose $s_1 = 2$ as a way to create 1000 kWh of electricity. This then leads to $s_2 = 2$ to supply the necessary fuel. But this also means that 20 kWh of electricity is used internally for producing fuel, hence we need to increase s_1 by another 2, in order to compensate for the electricity that is used inside the system. This on its turn exerts an extra fuel demand which leads to an increase of s_2 by 0.04. Then, this second round needs another 0.4 kWh of electricity, which leads to an increase of s_1 by 0.04. And so on, and so on. In the end, we obtain

$$\mathbf{s} = \begin{pmatrix} 100 + 2 + 0.04 + 0.0008 + \cdots \\ 2 + 0.04 + 0.0008 + \cdots \end{pmatrix} = \begin{pmatrix} 102.0408\cdots \\ 2.0408\cdots \end{pmatrix} \quad (4.24)$$

This sequential solution to a system of essentially simultaneous equations is tedious. It is slow, subject to errors, and there is no clear condition of when to stop iterating. One can use a fixed number of rounds, say 10, or use a criterion to stop further calculations when the relative change of the scaling factors in the last round decreases below a fixed threshold, say 10^{-6}.

The approach presented here can be said to delineate a network with recursive relationships as a linear sequence of infinite length. A process that is recursively called appears again and again, with ever-decreasing scaling factors. Asymptotically, it leads to the same solution as the exact approach of Section 4.3.1.

4.3.3   Algebraic manipulation of systems of equations

An alternative to the sequential approach is to algebraically manipulate the system of equations as in the following sequence of steps:

$$-2 \times s_1 + 100 \times s_2 = 0 \Leftrightarrow 2 \times s_1 = 100 \times s_2 \Leftrightarrow s_1 = 50 \times s_2 \qquad (4.25)$$

which be inserted into the second equation

$$10 \times 50 \times s_2 + -10 \times s_2 = 1000 \Leftrightarrow 490 \times s_2 = 1000 \Leftrightarrow s_2 = 2.0408 \cdots \quad (4.26)$$

This can the be used to find the solution for s_1:

$$s_1 = 50 \times 2.0408 \cdots = 102.0408 \cdots \qquad (4.27)$$

It will be clear that this algebraic manipulation is useless in large systems with 500 or 1000 equations.

4.3.4   The use of a power series expansion

In Section 4.2, the power series expansion of the inverse matrix was discussed as a computational shortcut that has become obsolete. It may serve a quite different purpose, related to providing an understanding of the perpetuating recurrence or processes in an infinitely large sequential system. The inverse of the technology matrix cannot be subject to series expansion, because the technology matrix does not meet the required conditions. However, keeping in mind the meaning of the co-ordinates, we can rearrange and manipulate **A** in a way such that the conditions are met and power series expansion is possible. First, we may observe that columns of the technology matrix may be interchanged, as long as we apply the same permutation to the rows of the scaling vector and to the columns of the intervention matrix. Next, we may notice that entire columns of the technology matrix may be multiplied with a constant, as long as we apply the same multiplication to the intervention matrix and divide the corresponding row of the scaling vector by this constant. We choose a rescaling such that the diagonal elements of the technology matrix become one. Thus, we end up with constructing a new technology matrix

$$\mathbf{A}' = \begin{pmatrix} 1 & -0.2 \\ -0.1 & 1 \end{pmatrix} \qquad (4.28)$$

For the power series expansion, we note that

$$(\mathbf{I} - \mathbf{A}') = \begin{pmatrix} 0 & 0.2 \\ 0.1 & 0 \end{pmatrix}, \quad (\mathbf{I} - \mathbf{A}')^2 = \begin{pmatrix} 0.02 & 0 \\ 0 & 0.02 \end{pmatrix},$$

$$(\mathbf{I} - \mathbf{A}')^3 = \begin{pmatrix} 0 & 0.004 \\ 0.002 & 0 \end{pmatrix}, \quad \text{etc.}$$

(4.29)

so that

$$\mathbf{I} + (\mathbf{I} - \mathbf{A}') + (\mathbf{I} - \mathbf{A}')^2 + (\mathbf{I} - \mathbf{A}')^3 = \begin{pmatrix} 1.02 & 0.204 \\ 0.102 & 1.02 \end{pmatrix} \quad (4.30)$$

When applied to the final demand vector, we find, up to 3rd order,

$$\mathbf{s}' \approx \begin{pmatrix} 1.02 & 0.204 \\ 0.102 & 1.02 \end{pmatrix} \begin{pmatrix} 0 \\ 1000 \end{pmatrix} = \begin{pmatrix} 204 \\ 1020 \end{pmatrix} \quad (4.31)$$

which is consistent with the previously found

$$\mathbf{s} = \begin{pmatrix} 102.0408 \cdots \\ 2.0404 \cdots \end{pmatrix} \quad (4.32)$$

provided that we take account of the interchanging of the rows of $\mathbf{s}'$ and the rescaling of one process by a factor of 10 and the other process by a factor of 100.

A more general form is as follows: the technology matrix and its transformed form are related by two operations: a permutation of columns and a rescaling of those columns. The permutation can be achieved by right multiplication with a matrix $\mathbf{K}$ which is itself a permuted form of the identity matrix:

$$\mathbf{AK} = \begin{pmatrix} -2 & 100 \\ 10 & -10 \end{pmatrix} \begin{pmatrix} 0 & 1 \\ 1 & 0 \end{pmatrix} = \begin{pmatrix} 100 & -2 \\ -10 & 10 \end{pmatrix} \quad (4.33)$$

and the rescaling proceeds by right multiplication with a diagonal matrix $\mathbf{L}$ with elements that are the reciprocal of the diagonal elements of the matrix that is to be transformed:

$$(\mathbf{AK})\mathbf{L} = \begin{pmatrix} 100 & -2 \\ -10 & 10 \end{pmatrix} \begin{pmatrix} 0.01 & 0 \\ 0 & 0.1 \end{pmatrix} = \begin{pmatrix} 1 & -0.2 \\ -0.1 & 1 \end{pmatrix} \quad (4.34)$$

Thus one has

$$\mathbf{A}' = \begin{pmatrix} 1 & -0.02 \\ -0.1 & 1 \end{pmatrix} =$$

$$\mathbf{AKL} = \begin{pmatrix} -2 & 100 \\ 10 & -10 \end{pmatrix} \begin{pmatrix} 0 & 1 \\ 1 & 0 \end{pmatrix} \begin{pmatrix} 0.01 & 0 \\ 0 & 0.1 \end{pmatrix}$$

(4.35)

The system of equations solved with the transformed technology matrix is

$$\mathbf{A}'\mathbf{s}' = \mathbf{f} \tag{4.36}$$

so that the untransformed scaling vector can be found by

$$\mathbf{s} = \mathbf{A}^{-1}\mathbf{f} = \mathbf{A}^{-1}\mathbf{A}'\mathbf{s}' = \mathbf{A}^{-1}\mathbf{A}\mathbf{K}\mathbf{L}\mathbf{s}' = \mathbf{K}\mathbf{L}\mathbf{s}' \tag{4.37}$$

We may also merge the permutation and rescaling matrices $\mathbf{K}$ and $\mathbf{L}$ into one transformation matrix $\mathbf{M}$. Then, transforming by

$$\mathbf{A}' = \mathbf{A}\mathbf{M} \tag{4.38}$$

leads to a solution which may be backtransformed by

$$\mathbf{s} = \mathbf{M}\mathbf{s}' \tag{4.39}$$

Hence, if we manage to find a matrix $\mathbf{M}$ which transforms the technology matrix into a matrix that fulfils the conditions discussed in Section 4.2, the use of a power series expansion to solve the equations is permissible.

The permutation and rescaling of processes enables one to employ the power series expansion after all, at least in the present case, and in most if not all cases. For reasons of computational efficiency, this result is no longer important with the advance of fast computers. But for reasons of understanding the convergence of the sequential model, it is still useful. We can express $\mathbf{s}$ in terms of a large number of contributions:

$$\mathbf{s} = \begin{pmatrix} 100 \\ 0 \end{pmatrix} + \begin{pmatrix} 0 \\ 2 \end{pmatrix} + \begin{pmatrix} 2 \\ 0 \end{pmatrix} + \begin{pmatrix} 0 \\ 0.04 \end{pmatrix} + \cdots \tag{4.40}$$

This clearly decomposes the scaling vector into a direct term plus several indirect terms (*cf.* Boustead & Hancock's (1979) discussion of "orders of analysis"). It also shows that, in the present case, the indirect contributions add slightly more than 2% to the direct term. In this way, one may use the power series expansion as a means to investigate the error introduced upon truncating an infinite sequence after a finite number of iterations.

4.4 Singularity of the technology matrix

In connection with the inversion of the technology matrix, it was remarked in Section 2.4 that the technology matrix must be square and non-singular. Several cases in which the squareness was violated have been discussed in

Chapter 3, for instance in relation to cut-off, multifunctionality and aggregated systems. It was shown that appropriate procedures for delivering a square technology matrix could be employed. Only for closed-loop recycling (Section 3.5) a rectangular matrix could be maintained, but even there, it was possible to construct a square matrix with an arbitrary allocation method on the basis of partitioning. However, the second requirement – non-singularity – was hardly discussed; only Section 3.5.4 provided one brief remark in that direction.

It is not possible to provide a general proof that the technology matrix, either the original one or the one that is constructed with procedures like allocation and cut-off, is non-singular. In concrete cases, one may check the determinant $\det(\mathbf{A})$: if it is 0 (or close to 0) the matrix is singular (or nearly singular). Even though a general proof is not available, there are good reasons to believe that a well-formed technology matrix is non-singular. For this, one should note several things.

First, each economic flow comes in one brand only. This means that there is only one process that produces a certain good or that treats a certain waste. If there are two process that make the same good, the goods are distinguished by origin (*e.g.*, beer brand X and beer brand Y), and a mixing process may be defined (see Section 3.4.3).

Second, it is not realistic that one process will need 2 litre of fuel to make 10 kWh of electricity, while another process will need to 5 kWh of electricity to make 1 litre of fuel. This presents an unfeasible economy, and a process which is so inefficient will disappear from the economic system. Still one must explicitly remove it from the technology matrix, as the matrix $\begin{pmatrix} -2 & 1 \\ 10 & -5 \end{pmatrix}$ is singular.

This example shows that a process which is a multiple of a different process (in this case, multiplied by $-1/2$) leads to a singular technology matrix. More complex cases in which one process is a linear combination of several other processes lead to the same problem of singularity. Columns are then dependent, and the matrix is said to be rank-deficient. Again, it is not likely that process from a feasible economy would form a singular technology matrix. One important exception to this is related to Monte Carlo simulations (see Section 6.4), where due to the stochastic nature of the coefficients in the technology matrix, realisations may occur in which the technology matrix is singular or close to singular.

4.5 Allocation in economic models

The discussion on allocation in connection to multifunctionality (Section 3.2) is not restricted to LCA. In fact there is a large literature on economic models in which these aspects show up as well. One problem is that these aspects are sometimes somewhat hidden, a second problem is that other terms then 'allocation' and 'substitution' are used, and that these terms are defined with a different meaning. The purpose of the present section is to explore in some detail how allocation enters these models. One additional complication is that these models often – but not exclusively – are based on the supply/use framework. It is therefore convenient to start by discussing this framework. Occasionally, we will meet the input-output framework. This is discussed in detail in Chapter 5.

4.5.1 The supply/use framework

The supply/use (or: make/use) framework strongly resembles the LCA framework, not in the sense of the ISO-framework (Section 1.2), but in the sense of the matrix structure, where rows denote flows and columns denote processes. The main difference is that LCA's technology matrix $\mathbf{A}$ is split into two matrices: the supply matrix $\mathbf{V}$ and the use matrix $\mathbf{U}$. The supply matrix only contains data with respect to outputs, while the use matrix concentrates on inputs. For certain reasons, rows of the supply matrix denote processes, while rows of the use matrix denote economic flows, just like LCA's technology matrix. To clarify the connection with LCA, we will therefore consistently use the transpose of $\mathbf{V}$, $\mathbf{V}^{\mathrm{T}}$.

In contrast to LCA, the inputs – the elements of the use matrix – are indicated by positive coefficients. Both the supply and the use matrix are positive semi-definite:

$$\mathbf{V}^{\mathrm{T}} \geq 0 \text{ and } \mathbf{U} \geq 0 \tag{4.41}$$

The relation between the two frameworks is simple:

$$\mathbf{A} = \mathbf{V}^{\mathrm{T}} - \mathbf{U} \tag{4.42}$$

Construction of $\mathbf{A}$ given $\mathbf{V}$ and $\mathbf{U}$ is trivial, and the only auxiliary information needed to carry out the opposing way of constructing of $\mathbf{V}$ and $\mathbf{U}$ given $\mathbf{A}$ is the fact that both matrices are positive semi-definite by definition.

Of interest is furthermore that activity analysis (Koopmans, 1951) is a field in economic analysis where the technology matrix is encountered (and even under that name, p.37) as a possibly rectangular matrix with similar features as in LCA. The main difference with LCA is that there is no 'one brand axiom,' and that the existence of more than one brand of many products creates a choice of suppliers. Additional constraints in the form of cost minimisation then lead to a linear-programming problem.

In the example system, given as

$$\mathbf{A} = \begin{pmatrix} -2 & 100 \\ 10 & 0 \end{pmatrix} \tag{4.43}$$

we find

$$\mathbf{U} = \begin{pmatrix} 2 & 0 \\ 0 & 0 \end{pmatrix} \text{ and } \mathbf{V}^{\mathrm{T}} = \begin{pmatrix} 0 & 100 \\ 10 & 0 \end{pmatrix} \tag{4.44}$$

As far as we know, the theory of supply/use analysis possesses no axiom that flows should come in one brand only. Hence, it is quite natural that the technology matrix (3.23), that was later onwards modified into (3.108), can occur in such an analysis. This then happens in the context of collecting economy-wide process data, not in an allocation step. Thus, the matrices

$$\mathbf{U} = \begin{pmatrix} 2 & 0 & 5 \\ 0 & 0 & 0 \\ 0 & 0 & 0 \end{pmatrix} \text{ and } \mathbf{V}^{\mathrm{T}} = \begin{pmatrix} 0 & 100 & 0 \\ 10 & 0 & 0 \\ 18 & 0 & 90 \end{pmatrix} \tag{4.45}$$

are in no sense suspicious. Note, however, that an LCA-perspective will deem the supply matrix as suspicious, because it is multifunctional, and it has one flow that comes in two brands.

Let us study what happens if we subject this system to the question of delivering an external demand

$$\mathbf{f} = \begin{pmatrix} 0 \\ 1000 \\ 0 \end{pmatrix} \tag{4.46}$$

representing 1000 kWh of electricity. This involves scaling factors $\mathbf{s}$, such that

$$\left(\mathbf{V}^{\mathrm{T}} - \mathbf{U}\right)\mathbf{s} = \mathbf{f} \tag{4.47}$$

This can be solved for $\mathbf{s}$ by means of

$$\mathbf{s} = \left(\mathbf{V}^{\mathrm{T}} - \mathbf{U}\right)^{-1}\mathbf{f} \tag{4.48}$$

Indeed, as $\mathbf{V}$ and $\mathbf{U}$ are square and $\mathbf{V}^T - \mathbf{U}$ is non-singular, inversion is possible and yields

$$\left(\mathbf{V}^T - \mathbf{U}\right)^{-1} = \begin{pmatrix} 0 & 0.1 & 0 \\ 0.01 & 0.001 & 0.000556 \\ 0 & -0.02 & 0.0111 \end{pmatrix} \quad (4.49)$$

so that

$$\mathbf{s} = \begin{pmatrix} 100 \\ 1 \\ -20 \end{pmatrix} \quad (4.50)$$

There are two noteworthy aspects of this solution:

- the solution shows that one process has a negative scaling factor, *i.e.* one process generates a negative output;

- the balance equation can be solved, apparently without an allocation step, even though the system contains a multifunctional process.

These two issues will be discussed briefly.

4.5.2 Negative scaling factors: the Hawkins-Simon condition

In the context of economic models, systems such as the above one have spawned extensive discussion. This seems to have started with the observation by Hawkins & Simon (1949) that a production system is "economically meaningful" (p.246) when their 'scaling factors' are all positive. Duchin & Szyld (1985, p.270) discuss the issue of negative scaling factors in a paper in which they present a "model with assured positive output." The interested reader is referred to Ten Raa *et al.* (1984) and the sequel by Ten Raa (1988) for an illustration of the problems involved in interpreting and discarding the negative numbers.

In the context of LCA, it is important to remember that the product system considered does not represent an economy-wide production system, but only that tiny part of an economy-wide production system that is connected with the functional unit, which is usually formulated in quite marginal terms. Therefore, the existence of negative scaling factors in the product system does not imply that certain processes are on the macro-level generating negative output, but only implies that these processes are producing slightly less than they would do when the product system is removed. To be more precise, one could think of a production system, characterised by a technology matrix $\mathbf{A}$, a final demand vector $\mathbf{f}_0$, and a scaling vector $\mathbf{s}_0$, connected by

$$\mathbf{A}\mathbf{s}_0 = \mathbf{f}_0 \tag{4.51}$$

In the change-oriented mode of LCA (Guinée et al, 2002), LCA can be interpreted as asking the question of the changes due to a small change in final demand $\mathbf{f}$ on top of the existing demand $\mathbf{f}_0$. This then leads to

$$\mathbf{A}\,(\mathbf{s}_0 + \mathbf{s}) = (\mathbf{f}_0 + \mathbf{f}) \tag{4.52}$$

which on its turn implies the well-known

$$\mathbf{A}\mathbf{s} = \mathbf{f} \tag{4.53}$$

As long as the changes are small, we will have $|s_j| \ll |s_{0j}|$, and the economy-wide scaling factors are still positive and meaningful. In short, product systems may have negative scaling factors, but production systems may not.

4.5.3 The technology assumptions

The second issue concerns the possibility to solve balance equations for a production system with multifunctional processes, without an allocation step. It is true that no recognisably allocation step has been made in solving the system depicted above. But one should observe the fact that in the case of LCA with substitution-based allocation (3.2.2) the unallocated technology matrix was to be written (see (3.108)) as

$$\mathbf{A}' = \begin{pmatrix} -2 & 100 & -5 \\ 10 & 0 & 0 \\ 18 & 0 & 0 \\ 0 & 0 & 90 \end{pmatrix} \tag{4.54}$$

which is not a solvable system. The allocation step consisted in an explicit declaration of the equivalence between the third and the fourth flow, in this case MJ of heat from facility X and MJ of heat from facility Y. After this step, the matrix

$$\mathbf{A}' = \begin{pmatrix} -2 & 100 & -5 \\ 10 & 0 & 0 \\ 18 & 0 & 90 \end{pmatrix} \tag{4.55}$$

resulted, a square and invertible matrix. The conclusion is that the system defined by (4.45) is obscured by the fact that the equivalence between heat from facility X and heat from facility Y has been built-in from the beginning. Denial of the 'no two brands' axiom thus means that hidden or implicit substitution-based allocation steps are made.

Of course, it also frequently happens that two brands are distinguished in a supply/use table. The system defined by

$$
\mathbf{U} = \begin{pmatrix} 2 & 0 & 5 \\ 0 & 0 & 0 \\ 0 & 0 & 0 \\ 0 & 0 & 0 \end{pmatrix} \text{ and } \mathbf{V}^{\mathrm{T}} = \begin{pmatrix} 0 & 100 & 0 \\ 10 & 0 & 0 \\ 18 & 0 & 0 \\ 0 & 0 & 90 \end{pmatrix} \tag{4.56}
$$

might also occur. In fact, the distinction of brands and multifunctionality is known to lead to rectangularity, and allocation-like procedures have been discussed at length in the theory of supply/use analysis (Konijn, 1994). Main schools are the following:

- the process-technology (or: industry-technology) model, in which every process is regarded as representing a unique technology to convert a bundle of inputs into a bundle of outputs;

- the commodity-technology model, in which every commodity (or: product) is assumed to be produced in a specific way, irrespective of the process where it is produced;

- the by-product technology model, in which a process's primary product is assumed to be produced in that particular way, and a process's non-primary product is assumed to be produced in the way according to the process that produces that product as its primary product.

The process-technology model does not solve the problem of rectangularity, unless a partitioning-like step is added. In many cases, this is done implicitly on an economic basis: outputs of co-products are all in monetary terms, so the inputs are distributed on the basis of the share in proceeds to the outputs. The commodity-technology model is clearly equivalent to the substitution method. The by-product technology model is a more sophisticated version of the commodity-technology model, along the lines that Weidema (2001) introduced in LCA.

Chapter 5

Relation with input-output analysis*

There is an interesting analogy between the technology matrix and the inventory problem on the one hand and an analytical tool for investigating industrial dependencies on the other hand. This latter type of analysis is called input-output analysis, or sometimes inter-industry analysis, and it has been introduced by Wassily Leontief around 1930. This chapter discusses the basic principles of input-output analysis (IOA) as originally introduced by Leontief, with an emphasis on its environmental extensions, and proceeds to discuss to important applications of IOA in relation to LCA: replacement of LCA by IOA, and combination of LCA and IOA. This chapter does not provide a comprehensive treatment of IOA; for this, the reader is referred to texts like Miller & Blair (1985) and Duchin & Steenge (1999).

5.1 Basics of input-output analysis and its environmental extension

In its original form, input-output analysis starts with a concept that is closely related to the technology matrix: the transactions matrix. This is a matrix of which a column represents the inputs of a process (or industry, or sector; although these terms occur frequently in the literature on IOA, we have here chosen to use the term process, to make the analogy with LCA as close as possible). But unlike the technology matrix, where the inputs of a process are formulated in terms of the flows of products (like fuel and electricity), the transactions matrix records the inputs in terms of the other processes' outputs (like fuel production and electricity production). Another difference is that the technology matrix contains a process' inputs and outputs, distinguished by the sign, while the transactions matrix contains a process' inputs only; minus signs are not needed. The outputs of a process is by definition one single type of output, not steel or electricity but a plain process' output. It is not written as a matrix element in the column that represents that process, but it can be figured out by aggregating all processes' inputs of that process' output. A final element to notice is that IOA has been developed to analyse the structure of the economy, with an emphasis on the flows of money. The coefficients of the transactions matrix thus measure the inputs of the steel-producing process in terms of euro, dollar or yen, not in physical units like kg, kWh or km. Occasionally, one sees approaches in which the transactions matrix is measured in physical terms.

Let us for instance consider the transaction matrix

$$\mathbf{Z} = \begin{pmatrix} 2 & 10 \\ 4 & 2 \end{pmatrix} \tag{5.1}$$

in a linear space where the first row and the first column denote the process of production of fuel and the second row and the second column the process of production of electricity. The meaning of element z_{12} is that the second process needs 10 euro of the first process' output. It also needs 2 euro of its own output (z_{22}). In contrast to the technology matrix, the transactions matrix $\mathbf{Z}$ is square by definition, as the rows and the columns refer to processes (sectors, industries).

To the transaction matrix $\mathbf{Z}$ that indicates the inter-industry demands, we may append the demand by households as a vector $\mathbf{y}$. In the example below, the households exert a demand of 8 euro of the first process' output and 4 euro of the second process' output:

$$(\, \mathbf{Z} \, | \, \mathbf{y} \,) = \left(\begin{array}{cc|c} 2 & 10 & 8 \\ 4 & 2 & 4 \end{array} \right) \tag{5.2}$$

Thus, the first process' total output is $2 + 10 + 8 = 20$ euro, and the second process' total output $4 + 2 + 4 = 10$ euro. It is convenient to define a vector $\mathbf{x}$ to indicate the total output:

$$(\, \mathbf{Z} \, | \, \mathbf{y} \, | \, \mathbf{x} \,) = \left(\begin{array}{cc|c|c} 2 & 10 & 8 & 20 \\ 4 & 2 & 4 & 10 \end{array} \right) \tag{5.3}$$

A next step in input-output analysis is the analysis of the economy-wide consequences of changes in household demand. For instance, what happens when the demand of 8 euro of the first process' output is increased to 28 euro? This means that the first process' output is increased from 20 to 40 euro. Under the assumption of linear scaling, its input of the second process' output is then doubled from 4 to 8 euro. This on its turn implies an increased production volume of the second process, with a subsequent increase of this process' input of electricity. And so on.

It is convenient to define a matrix of technical coefficients $\tilde{\mathbf{Z}}$ by expressing each process' inputs as a fraction of its output

$$\tilde{\mathbf{Z}} = \begin{pmatrix} 2/20 & 10/10 \\ 4/20 & 2/10 \end{pmatrix} = \begin{pmatrix} 0.1 & 1 \\ 0.2 & 0.2 \end{pmatrix} \tag{5.4}$$

or more generally

$$\tilde{\mathbf{Z}} = \mathbf{Z} \left(\mathbf{diag}(\mathbf{x}) \right)^{-1} \tag{5.5}$$

where $\mathbf{diag}(\mathbf{x})$ is the matrix that consists of the elements of $\mathbf{x}$ at the diagonal and zeros at all off-diagonal places. The question is then to express the processes' new total output vector $\mathbf{x}'$ as a function of the new households demand vector $\mathbf{y}'$. The expression is

$$\mathbf{x}' = \left(\mathbf{I} - \tilde{\mathbf{Z}} \right)^{-1} \mathbf{y}' \tag{5.6}$$

where $\mathbf{I}$ is the identity matrix of the same size as $\tilde{\mathbf{Z}}$. Here it is a 2×2-matrix:

$$\mathbf{I} = \begin{pmatrix} 1 & 0 \\ 0 & 1 \end{pmatrix} \tag{5.7}$$

Notice that we must distinguish here between the total output and households demand vectors $\mathbf{x}$ and $\mathbf{y}$ in the existing situation, and the same vectors $\mathbf{x}'$ and $\mathbf{y}'$ in a new situation. This allows us to calculate the effects of a change $\Delta \mathbf{y} = \mathbf{y}' - \mathbf{y}$ on $\Delta \mathbf{x} = \mathbf{x}' - \mathbf{x}$:

$$\Delta \mathbf{x} = \left(\mathbf{I} - \tilde{\mathbf{Z}} \right)^{-1} \Delta \mathbf{y} \tag{5.8}$$

Without loss of generality, we may take the starting point as a reference situation, thus putting $\mathbf{x} = \mathbf{0}$, so that $\mathbf{y}'$ coincides with $\Delta \mathbf{y}$. Inserting the unprimed symbols then yields the most usual form:

$$\mathbf{x} = \left(\mathbf{I} - \tilde{\mathbf{Z}} \right)^{-1} \mathbf{y} \tag{5.9}$$

where the reinterpretation of $\mathbf{y}$ as the households demand vector and $\mathbf{x}$ as the total output vector can be made.

The term $\left(\mathbf{I} - \tilde{\mathbf{Z}}\right)^{-1}$ is known as the Leontief inverse. In the example, it is

$$\left(\mathbf{I} - \tilde{\mathbf{Z}}\right)^{-1} = \begin{pmatrix} 1.54 & 1.92 \\ 0.38 & 1.73 \end{pmatrix} \tag{5.10}$$

For a households demand

$$\mathbf{y} = \begin{pmatrix} 28 \\ 4 \end{pmatrix} \tag{5.11}$$

one then finds

$$\mathbf{x} = \begin{pmatrix} 50.78 \\ 17.69 \end{pmatrix} \tag{5.12}$$

for the processes' total output vector. Because $\mathbf{Z}$ is square, the expression $\mathbf{I} - \tilde{\mathbf{Z}}$ represents a square matrix as well. Although it may still be singular in exceptional cases, there is no allocation step needed to transform a rectangular matrix into a square one as in LCA.

In LCA, the final demand vector normally consists of zeros at all but one place. Suppose that we in IOA make the special choice of zero households demand for all processes but one, and for that process a demand of 1 euro. Using this for $\mathbf{y}$, the resulting vector $\mathbf{x}$ measures the output of all processes as a result of 1 euro demand from one particular process. The elements of $\mathbf{x}$ are known as the Leontief multipliers.

Although primarily designed as a tool for the analysis of economic dependencies, IOA can serve environmental analyses as well when it is extended with so-called satellite accounts for the environmental flows. Below the transactions matrix, additional rows to account for carbon dioxide, sulphur dioxide, crude oil, etc. are added. The satellite matrix $\tilde{\mathbf{B}}$ is structurally identical to the intervention matrix $\mathbf{B}$ of the inventory analysis. Environmental interventions associated with a certain households demand vector $\mathbf{y}$ are then

$$\mathbf{g} = \tilde{\mathbf{B}} \left(\mathbf{I} - \tilde{\mathbf{Z}}\right)^{-1} \mathbf{y} \tag{5.13}$$

Some care should be taken in directly comparing $\mathbf{g}$ from LCA with from IOA, as well as using $\mathbf{B}$ from LCA as $\tilde{\mathbf{B}}$ in IOA or the other way around, as there may be differences in the ordering or level of aggregation. Moreover, $\tilde{\mathbf{B}}$ is defined per euro of process output, while $\mathbf{B}$ is, like $\mathbf{A}$, defined with respect to an arbitrary reference unit, such as $100,000$ kg of steel or 1000 TV sets.

5.2 Comparison of LCA and IOA

The Leontief inverse plays a role that is comparable to the inverse of the technology matrix; see Table 5.1 for an overview of analogous concepts.

Table 5.1: Overview of analogous concepts in life cycle inventory analysis and input-output analysis.

life cycle inventory analysis	LCA	input-output analysis	IOA
product, economic flow		commodity	
(unit) process		industry, sector, establishment	
		transactions matrix	$\mathbf{Z}$
technology matrix	$\mathbf{A}$	technical coefficients matrix	$\tilde{\mathbf{Z}}$
final demand vector	$\mathbf{f}$	households demand vector	$\mathbf{y}$
scaling vector	$\mathbf{s}$	total output vector	$\mathbf{x}$
inverse of technology matrix	$\mathbf{A}^{-1}$	Leontief inverse	$\left(\mathbf{I} - \tilde{\mathbf{Z}}\right)^{-1}$
intervention matrix	$\mathbf{B}$	satellite matrix	$\tilde{\mathbf{B}}$
equation for $\mathbf{s}$	$\mathbf{s} = \mathbf{A}^{-1}\mathbf{f}$	equation for $\mathbf{x}$	$\mathbf{x} = \left(\mathbf{I} - \tilde{\mathbf{Z}}\right)^{-1}\mathbf{y}$
equation for $\mathbf{g}$	$\mathbf{g} = \mathbf{B}\mathbf{s}$	equation for $\mathbf{g}$	$\mathbf{g} = \tilde{\mathbf{B}}\mathbf{x}$

Despite the similarities between LCA and IOA, there are some subtle differences. These ultimately relate to the difference in set-up of the matrices. The transactions matrix $\mathbf{Z}$ of IOA is of the type process×process (the most common terms for this are industry×industry and sector×sector). This means that the labels of rows and columns refer to processes (or industries, or sectors). The technology matrix $\mathbf{A}$ of LCA is of the type commodity×process, so that the labels of rows refer to commodities (or products, or economic flows) while the labels headers of columns refer to process (or industries, or sectors). Some important consequences are the following:

- The transactions matrix of IOA specifies the flow from one process to another process, without paying regard to the product that actually flows between these two processes. In LCA, the technology matrix specifies the flow of products to and from a process, without an explicit specification of the origin or destination of such flows.

- As we have seen in Section 3.2, an important problem in LCA is the fact that the technology matrix is often not square, so that matrix inversion cannot be applied, at least not directly. Allocation procedures must then be employed. The transactions matrix of IOA is always square, and the Leontief inverse can almost always be calculated.

- The diagonal of a transactions matrix has a special meaning: it represents the inputs of an industry to itself. An example could be electric power plants that use a part of their own electricity, for instance for pumping fuel into the combustors, and agriculture, where corn produced is partly used to feed cattle. In LCA the diagonal has no particular meaning. In fact, it is often not tractable, because the technology matrix is rectangular. Moreover, changing the ordering of industries in IOA has no influence on the interpretation of the diagonal as self-inputs as long as the order of the rows and columns is the same. In LCA, the order of processes and products may be changed independently, so that any suggestion of a special diagonal (as in Frischknecht & Kolm (1995)) is spurious. As a side remark, it may be noted that the interpretation of diagonal elements in IOA in terms of self-inputs is questionable as well; see Waugh (1950), who writes that the transactions matrix is a "hollow" matrix, *i.e.* with zero diagonal, Edey & Peacock (1959) who write down a "—" in the diagonal entries, and Georgescu-Roegen (1971, p.256 *ff.*) who devotes an entire discussion to the subject.

- The coefficients of a technical coefficients matrix have a limited range of 'intelligible' values: they lie between 0 and 1, and the sum over one column does not exceed 1. For technical coefficients matrices in physical units and for LCA's technology matrix, these restrictions do not hold or are weaker. However, as the majority of texts on IOA deal with monetary units, certain properties and theorems (see, *e.g.*, Section 4.2) cannot be applied to IOA in physical terms and LCA.

- The conventional degree of specificity differs to quite some extent between LCA and IOA. In IOA, the entire economy is categorised into a few hundred industries, covering many different processes, with typical names as "metal ores mining." In LCA, a much finer categorisation is attempted; distinguishing mining of copper ore, iron ore, bauxite, etc.

One can say that IOA on the basis of an process×process table contains no information on commodities. This is often not entirely clear from texts dealing with IOA. Thus, one often sees expressions like "the industry plastics materials," where the label "the industry plastics materials production" would be more appropriate. Moreover, one should recognise that the households demand vector is not a demand of commodities (products), but a demand for industry's output. Demand is specified as "2 euro of plastics materials output," not as "2 euro of plastics materials."

As said, the exposition of IOA and the comparison with LCA are based on the original formulation of IOA. Original means here: on the basis of a process×process (or industry×industry) matrix. In the course of time modifications to this original scheme have been proposed, for instance leading to IOA on the basis of a commodity×process (or commodity×industry) matrix. For the purpose of the present book, we have chosen to categorise such formats as LCA, supply/use framework, or activity analysis (see Konijn (1994)). Here, the typical feature of IOA is regarded to be its process×process structure, leading to a square matrix by definition and the characteristic Leontief inverse with the $\mathbf{I} - \tilde{\mathbf{Z}}$.

5.3 IOA instead of LCA

A number of authors (*e.g.*, Lave *et al.* (1995), Hendrickson *et al.* (1998), Joshi (2000)) criticise LCA for being incomplete and approach the problem of finding environmental interventions associated with a certain external demand by switching to IOA. An important problem of traditional LCA is that process data must be collected for a very large number of processes, and that cut-offs (see Section 3.1) are needed at many places. Several decades of institutionalised compilation of IO-tables have resulted in a fairly complete and accurate picture of inter-industry flows. Thus, attempts have been made to replace the technology matrix by the input-output matrix. The approach is known as "economic input-output life-cycle analysis" or EIO-LCA in short.

Instead of the process-specific environmental coefficients, national emission tables like the US Toxics Release Inventory are coupled to the IO-matrix. Thus

$$\mathbf{g} = \mathbf{B}\mathbf{A}^{-1}\mathbf{f} \qquad (5.14)$$

is replaced by

$$\mathbf{g} = \tilde{\mathbf{B}}\left(\mathbf{I} - \tilde{\mathbf{Z}}\right)^{-1}\mathbf{y} \qquad (5.15)$$

Lenzen (2001) compares "conventional and input-output-based life-cycle inventories" and concludes that errors due to cut-off may be larger than the errors introduced by using IOA. Lave *et al.* (1995) show that indirect discharges by computer production exceed direct discharges by a factor of 26. On the other hand, several authors mentioned discuss limitations (see, e.g., Nielsen & Pedersen Weidema (2001)). These include:

- the input-output tables themselves do not cover the entire life cycle; most prominently consumption processes and waste treatment are excluded;

- national IO-tables have separate entries for import and export, and hence tend to exclude interventions from production abroad;

- the industry classification of IO-tables is quite coarse, so that aggregation errors will be introduced.

In addition to the latter fact, we would like to point out that LCA is very often used to compare fairly similar products, such as bottles from polyethylene and bottles from polycarbonate, or in product design situations, where a designer wants to know the effects of changing materials. When these two materials are lumped together in one classification as "plastic materials," the range of applications of LCA is seriously restricted. The crucial connecting element is that we must translate a reference flow in the final demand vector $\mathbf{f}$ into a households demand vector $\mathbf{y}$. Whenever this is not possible, EIO-LCA is doomed to give up.

Several authors in the field of energy analysis discuss the use of input-output analysis, sometimes as opposed to the type of analysis that traditionally shows up in LCA, which is then called "process analysis" or "vertical analysis." Boustead & Hancock (1979) and IFIAS (1974) are standard references in this respect. Miller & Blair (1985) also discuss "energy input-output analysis" and "environmental input-output analysis." And pathbreaking studies on the relation between economic chains and the environment, such as Ayres & Kneese (1969) and Leontief (1970) are based on (process×process) IOA as well, as are newer applications, such as Perrings (1987). An exception that should be mentioned here is Victor (1972), who bases an economy-wide study on the commodity×process structure.

5.4   Hybrid analysis

As discussed above, LCA yields quite specific data but suffers from providing an incomplete picture. IOA, in contrast, yields a fairly complete system, but is in certain respects overly aggregated and hence unspecific. The result is for both approaches an increase of uncertainty. This observation leaded various efforts to combine the strengths of both, which are generally called hybrid analysis. In general, the IOA-based part in a tiered hybrid analysis provides relatively complete far upstream system boundaries while the LCA-based part provides a much more specific near upstream system and the downstream boundaries (see Marheineke *et al.* (1998)).

For the purpose of analysis, we discuss in the next two sections two different approaches here under the names tiered hybrid analysis and internally solved hybrid analysis. The reader should acknowledge, however, that the connection of LCA and IOA in a hybrid analysis is a topic to which current much research is devoted. For a more comprehensive presentation, see Suh & Huppes (2002).

5.4.1 Tiered hybrid analysis

The concept of hybrid analysis appears from the 1970s in energy analysis. Bullard *et al.* (1978), Van Engelenburg *et al.* (1994) and Treloar (1997) combined process analysis based on flow charts with IOA to calculate net energy requirements of a product or an entire economy. The computational structure of tiered hybrid analysis can be formulated from the following phrases (Bullard *et al.*, 1978, p.281–282): "Some of the input materials may be typical products of I-O sectors ... Thus the only input materials requiring further process analysis are atypical products not easily classified in an I-O sector." It is remarkable that, like in LCA, process analysis is not described in mathematical terms while input-output analysis is. A mathematical interpretation would be as follows:

$$\mathbf{g} = \mathbf{g}_{\mathrm{IOA}} + \mathbf{g}_{\mathrm{LCA}} = \tilde{\mathbf{B}}\left(\mathbf{I} - \tilde{\mathbf{Z}}\right)^{-1}\mathbf{y} + \mathbf{B}\mathbf{A}^{-1}\mathbf{f} \qquad (5.16)$$

Thus, $\mathbf{y}$ determines the IOA-based part of the hybrid analysis, while $\mathbf{f}$ determines the LCA-based part. There is no interaction between $\mathbf{y}$ and $\mathbf{f}$, nor between $\tilde{\mathbf{Z}}$ and $\mathbf{A}$; this accounts for the qualification 'tiered.' It is up to the user to specify the separation between the IOA-based and the LCA-based demand in an appropriate way. Note that we may symbolically unify the two tiers as

$$\mathbf{g} = \begin{pmatrix} \tilde{\mathbf{B}} & \mathbf{B} \end{pmatrix} \begin{pmatrix} \mathbf{I} - \tilde{\mathbf{Z}} & \mathbf{0} \\ \mathbf{0} & \mathbf{A} \end{pmatrix}^{-1} \begin{pmatrix} \mathbf{y} \\ \mathbf{f} \end{pmatrix} \qquad (5.17)$$

so that the general form employed in this book may be maintained (Suh & Huppes, 2001).

This presents just one approach towards partitioning the final demand into a commodity part, generated with an LCA-system, and a process part, generated with an IOA-system. There are two radically different choices possible for this partitioning. In LCA, we have seen the following choices (notice that the first one is inspired by the above mentioned energy analysis):

- Processes are in principle modeled in the IOA-part, but processes not covered by the input-output table, mainly consumption and waste treatment, are modeled in the LCA-part (Moriguchi *et al.* (1993), Hondo *et al.* (1996), Joshi (2000)).

- Processes around the production and consumption stage are modeled in the LCA-part (as "foreground" processes), and processes further upstream and downstream in the IOA-part (as "background" processes). This approach is taken by Suh & Huppes (2002) and Guinée *et al.* (2002), and forms the idea behind the MIET software.

It is interesting that these two strategies point towards several defects of using IOA for replacing LCA altogether:

- IO-tables in general do not comprise activities related to consumption and waste treatment;

- IO-tables are not very detailed with respect to specific modes of production;

- A national IO-table does not provide a complete upstream system boundary if the national economy is severely dependent upon imported goods.

We see that IOA may be supplemented by LCA, or that LCA may be supplemented by IOA. As a natural choice for this book, we will concentrate on the latter form. Hence, it is logical to rewrite (5.17) as follows:

$$\mathbf{g} = (\mathbf{B} \quad \tilde{\mathbf{B}}) \begin{pmatrix} \mathbf{A} & \mathbf{0} \\ \mathbf{0} & \mathbf{I} - \tilde{\mathbf{Z}} \end{pmatrix}^{-1} \begin{pmatrix} \mathbf{f} \\ \mathbf{y} \end{pmatrix} \qquad (5.18)$$

We will elaborate this form below and in the next section.

The term 'tiered' is often understood to mean another thing as well, namely the case of an LCA-based core with IOA-based peripherals (or the other way around). This requires, however, that the vector $\mathbf{y}$ is $\mathbf{0}$ with regard to final demand, but not with respect to internal delivery to the LCA-based part, for which $\mathbf{f}$ is the final demand vector. In other words, $\mathbf{f}$ determines $\mathbf{s}$, and this in turn determines $\mathbf{x}$. To formalise such a procedure, we will develop a simple example.

Consider the technology matrix below where the last two rows refer to the use of generators (in euro) and the use of refineries (in euro):

$$\mathbf{A} = \begin{pmatrix} -2 & 100 \\ 10 & 0 \\ -0.1 & 0 \\ 0 & -1 \end{pmatrix} \tag{5.19}$$

That is, for producing 10 kWh of electricity, one needs 2 litre of fuel and 0.1 euro of generators, for instance in the form of depreciation. Assume that an IO-table is known for these two sectors:

$$\tilde{\mathbf{Z}} = \begin{pmatrix} 0.1 & 0.5 \\ 0.2 & 0.2 \end{pmatrix} \tag{5.20}$$

indicating the economic dependency of the generator-producing industry and refineries.

One partitions the technology matrix into a part that is dealt with in the traditional LCA-way, $\mathbf{A}'$, and a part that is dealt with by IOA, $\mathbf{A}''$:

$$\mathbf{A} = \left(\frac{\mathbf{A}'}{\mathbf{A}''} \right) = \begin{pmatrix} -2 & 100 \\ 10 & 0 \\ \hline -0.1 & 0 \\ 0 & -1 \end{pmatrix} \tag{5.21}$$

A similar partitioning of the final demand vector takes place:

$$\mathbf{f} = \left(\frac{\mathbf{f}'}{\mathbf{f}''} \right) = \begin{pmatrix} 0 \\ 1000 \\ \hline 0 \\ 0 \end{pmatrix} \tag{5.22}$$

It is clear that the LCA-part of the final demand vector $\mathbf{f}' = \begin{pmatrix} 0 & 1000 \end{pmatrix}^{\mathrm{T}}$ leads to scaling factors

$$\mathbf{s} = (\mathbf{A}')^{-1} \mathbf{f}' = \begin{pmatrix} -2 & 100 \\ 10 & 0 \end{pmatrix}^{-1} \begin{pmatrix} 0 \\ 1000 \end{pmatrix} = \begin{pmatrix} 100 \\ 2 \end{pmatrix} \tag{5.23}$$

The scaling factors also apply to the IOA-part given by $\mathbf{A}''$:

$$\mathbf{A}'' \mathbf{s} = \begin{pmatrix} -0.1 & 0 \\ 0 & -1 \end{pmatrix} \begin{pmatrix} 100 \\ 2 \end{pmatrix} = \begin{pmatrix} -10 \\ -2 \end{pmatrix} = \mathbf{f}'' \tag{5.24}$$

This latter vector $\mathbf{f}''$ can be interpreted as the LCA-parts requirement to the IOA-part of the system. Adhering to the balancing principle, the IOA-part must satisfy it. Hence, the IOA-part must deliver $-\mathbf{f}''$ in the form of a households demand vector $\mathbf{y}$; notice that we are allowed to establish the identity

$$\mathbf{y} = -\mathbf{f}'' \tag{5.25}$$

by virtue of the fact that $\mathbf{f}''$ is formulated in monetary terms, and can thus be interpreted as a process output instead of the commodity itself.

Now we find the subsequent IOA-problem:

$$\left(\mathbf{I} - \tilde{\mathbf{Z}}\right)^{-1}\mathbf{y} = \begin{pmatrix} 0.9 & -0.5 \\ -0.2 & 0.8 \end{pmatrix}^{-1} \begin{pmatrix} 10 \\ 2 \end{pmatrix} = \begin{pmatrix} 14.5 \\ 6.1 \end{pmatrix} = \mathbf{x} \tag{5.26}$$

so that the total output vector $\mathbf{x}$ is found.

When the intervention matrices $\mathbf{B}$ and $\tilde{\mathbf{B}}$ are available for the LCA and IOA system respectively, and assuming that the ordering of the rows of these two matrices is identical, we have

$$\mathbf{g}_{\text{LCA}} = \mathbf{Bs} \tag{5.27}$$

and

$$\mathbf{g}_{\text{IOA}} = \tilde{\mathbf{B}}\mathbf{x} = -\tilde{\mathbf{B}}\left(\mathbf{I} - \tilde{\mathbf{Z}}\right)^{-1}\mathbf{A}''\mathbf{s} \tag{5.28}$$

which yield together

$$\mathbf{g} = \mathbf{g}_{\text{LCA}} + \mathbf{g}_{\text{IOA}} = \left(\mathbf{B} - \tilde{\mathbf{B}}\left(\mathbf{I} - \tilde{\mathbf{Z}}\right)^{-1}\mathbf{A}''\right)\left(\mathbf{A}'\right)^{-1}\mathbf{f}' \tag{5.29}$$

This form can also be expressed in a form like (5.18):

$$\mathbf{g} = \begin{pmatrix} \mathbf{B} & \tilde{\mathbf{B}} \end{pmatrix} \begin{pmatrix} \mathbf{A}' & \mathbf{0} \\ \mathbf{A}'' & \mathbf{I} - \tilde{\mathbf{Z}} \end{pmatrix}^{-1} \begin{pmatrix} \mathbf{f} \\ \mathbf{0} \end{pmatrix} \tag{5.30}$$

5.4.2   Internally solved hybrid analysis

Recently another hybrid approach has been developed, which is called LCA based hybrid analysis (Suh & Huppes, 2000), but which we will here contrast with the tiered hybrid analysis by naming it the internally solved hybrid analysis. In this approach the technology matrix of a product system in LCA is fully interlinked with technical coefficients matrix of an economy in IOA. The form is

$$\mathbf{g} = (\; \mathbf{B} \quad \tilde{\mathbf{B}} \;) \begin{pmatrix} \mathbf{A}' & \mathbf{A}''' \\ \mathbf{A}'' & \mathbf{I} - \tilde{\mathbf{Z}} \end{pmatrix}^{-1} \begin{pmatrix} \mathbf{f} \\ \mathbf{0} \end{pmatrix} \qquad (5.31)$$

where once more $\mathbf{A}''$ represents the dependence of the LCA-system on the IOA-system, but now there is a dependence of the IOA-system on the LCA-system as well, embodied by $\mathbf{A}'''$. Interestingly, we may still pursue to use the expansion of the inverse of a partitioned matrix (see the Appendix, Section A.5) and express the above as the sum of a pure LCA-based part and an additional part due to the two-sided interaction with the IOA-system. It should be noticed, however, that the additional part contains terms that depend on $\mathbf{A}'$, so that the LCA- and IOA-parts are not truly separable.

Chapter 6

Perturbation theory

There is an extensive theoretical literature on the influence of perturbation of coefficients of matrices on solutions of systems of equations; see for instance Atkinson (1989) and in particular Stewart & Sun (1990). This theory can be used for a number of interesting subjects in LCA. First, the coefficients that describe the unit processes and define the technology matrix and the intervention matrix often suffer from uncertainty. A statistical treatment of the propagation of these uncertainties into uncertainties of scaling factors or environmental interventions is obviously important. The issue of uncertainties in LCA has been addressed by many authors. Here, we will restrict the discussion to generally applicable treatments, such as Heijungs (1994, 1996), Huijbregts (1998), Roš (1998) and Ciroth (2001), but the text presents many insights that have not been discussed within LCA before. Second, approaches have been developed under the name of marginal analysis or perturbation analysis (Heijungs, 1994) to investigate options for product improvement by means of perturbation-theoretic considerations. Finally, the influence of round-off in data and limited computer accuracy can also be addressed with perturbation-theoretic concepts. As far as we know, this issue has not been addressed before within the LCA-literature.

It should be mentioned that perturbation theory has been discussed in relation to input-output analysis (see also Section 5.1). For instance, Sebald (1974) and Van Dijk & Sladký (1995) provide interesting treatments. The direct applicability of such approaches is limited, however, because the matrices encountered in input-output analysis have characteristics that are not shared by the matrices that are used in LCA. For instance, matrices for input-output analysis are in most cases positive definite, and have column sums that are bounded below 1. These properties are employed in the mathematical proofs, but cannot be assumed for LCA. Therefore, these texts can only serve a role as suggesting approaches, but their factual results cannot be used directly. As a consequence, the theoretical level of this chapter is higher than that of the other chapters. Besides Heijungs (1994), the only place in which perturbation theory is discussed in relation to LCA and of which we are aware is a short contribution by Halada *et al.* (1998).

6.1   Some general results

Perturbation theory studies the influence of perturbations of coefficients of equations on the solutions to those equations. As a concrete example, consider the old example

$$\mathbf{As} = \mathbf{f} \tag{6.1}$$

with

$$\mathbf{A} = \begin{pmatrix} -2 & 100 \\ 10 & 0 \end{pmatrix} \tag{6.2}$$

and

$$\mathbf{f} = \begin{pmatrix} 0 \\ 1000 \end{pmatrix} \tag{6.3}$$

which is known to lead to

$$\mathbf{s} = \begin{pmatrix} 100 \\ 2 \end{pmatrix} \tag{6.4}$$

A typical question addressed in perturbation theory is as follows: given the above system, what would the solution be of the perturbed system with

$$\mathbf{A}' = \begin{pmatrix} -3 & 100 \\ 10 & 0 \end{pmatrix} \tag{6.5}$$

Obviously, one could use matrix inversion on the perturbed technology matrix to find the new solution. But it is in some respects more interesting to study how the perturbation propagates in the system, and to express the new solution in terms of the old one. For this, we will consider the perturbed matrix to consist of the original matrix plus a perturbation term:

$$\mathbf{A}' = \mathbf{A} + \delta\mathbf{A} \tag{6.6}$$

with in this case

$$\delta\mathbf{A} = \begin{pmatrix} -1 & 0 \\ 0 & 0 \end{pmatrix} \tag{6.7}$$

and try to derive expression for $\delta\mathbf{s}$, defined by

$$\mathbf{s}' = \mathbf{s} + \delta\mathbf{s} \tag{6.8}$$

These expressions will then be in terms of the original system and the perturbation term $\delta\mathbf{A}$. We will specifically address three forms of perturbations:

- those in which the perturbation is arbitrarily large or small, so that exact expressions must be employed;

- those in which the perturbation is small, so that differential calculus may be applied as a first order approximation;

- those in which the perturbation is stochastic, so that probability theory may be applied.

We will first develop exact expressions, which will later onwards be subject to simplifications for the special cases of small and random perturbations.

6.1.1 Perturbation of the technology matrix

Let us start from a square invertible technology matrix $\mathbf{A}$. Now assume that a change is introduced in this technology matrix. We will indicate the perturbation introduced by the matrix $\delta\mathbf{A}$. Thus, we will study the technology matrix $\mathbf{A} + \delta\mathbf{A}$. We will study what the effects of such a change are on the scaling vector $\mathbf{s}$. The change in scaling factors will be denoted as $\delta\mathbf{s}$. We will assume that the final demand vector $\mathbf{f}$ remains unchanged. Thus, starting from

$$\mathbf{As} = \mathbf{f} \tag{6.9}$$

we will study

$$(\mathbf{A} + \delta\mathbf{A})(\mathbf{s} + \delta\mathbf{s}) = \mathbf{f} \tag{6.10}$$

and in particular derive expressions for $\delta\mathbf{s}$. We proceed by expanding the terms in parentheses:

$$\mathbf{As} + \mathbf{A}\delta\mathbf{s} + \delta\mathbf{As} + \delta\mathbf{A}\delta\mathbf{s} = \mathbf{f} \tag{6.11}$$

By virtue of $\mathbf{As} = \mathbf{f}$, this reduces to

$$\mathbf{A}\delta\mathbf{s} + \delta\mathbf{As} + \delta\mathbf{A}\delta\mathbf{s} = \mathbf{0} \tag{6.12}$$

This can be rearranged into

$$(\mathbf{A} + \delta\mathbf{A})\delta\mathbf{s} = -\delta\mathbf{As} \tag{6.13}$$

or, assuming that $\mathbf{A} + \delta\mathbf{A}$ is invertible, into

$$\delta\mathbf{s} = -\left(\mathbf{A} + \delta\mathbf{A}\right)^{-1}\delta\mathbf{A}\mathbf{s} \tag{6.14}$$

Because $\mathbf{s} = \mathbf{A}^{-1}\mathbf{f}$, we finally obtain

$$\delta\mathbf{s} = -\left(\mathbf{A} + \delta\mathbf{A}\right)^{-1}\delta\mathbf{A}\mathbf{A}^{-1}\mathbf{f} \tag{6.15}$$

This is an important relationship. It enables us to explore the study of perturbation theory and statistical analysis of the inventory analysis.

Under the assumption that $\delta\mathbf{A}$ is infinitesimally small, we may approximate $\mathbf{A} + \delta\mathbf{A}$ by $\mathbf{A}$ and write

$$\frac{\partial\mathbf{s}}{\partial a_{ij}} = -\mathbf{A}^{-1}\frac{\partial\mathbf{A}}{\partial a_{ij}}\mathbf{A}^{-1}\mathbf{f} \tag{6.16}$$

This is in agreement with the differentiation rule for inverse matrices (see, *e.g.*, Harville (1997) and Balestra (1976)):

$$\frac{\partial\left(\mathbf{A}^{-1}\right)}{\partial a_{ij}} = -\mathbf{A}^{-1}\frac{\partial\mathbf{A}}{\partial a_{ij}}\mathbf{A}^{-1} \tag{6.17}$$

which can be easily connected to the previous formula through

$$\frac{\partial\mathbf{s}}{\partial a_{ij}} = \frac{\partial\left(\mathbf{A}^{-1}\right)}{\partial a_{ij}}\mathbf{f} + \mathbf{A}^{-1}\frac{\partial\mathbf{f}}{\partial a_{ij}} = \frac{\partial\left(\mathbf{A}^{-1}\right)}{\partial a_{ij}}\mathbf{f} \tag{6.18}$$

Individual coefficients of the vector $\dfrac{\partial\mathbf{s}}{\partial a_{ij}}$ depend on individual elements of $\dfrac{\partial\left(\mathbf{A}^{-1}\right)}{\partial a_{ij}}$. These are

$$\frac{\partial\left(\mathbf{A}^{-1}\right)_{mn}}{\partial a_{ij}} = -\left(\mathbf{A}^{-1}\right)_{mi}\left(\mathbf{A}^{-1}\right)_{jn} \tag{6.19}$$

This then yields

$$\frac{\partial s_k}{\partial a_{ij}} = \sum_l \frac{\partial\left(\mathbf{A}^{-1}\right)_{kl}}{\partial a_{ij}}f_l = \sum_l -\left(\mathbf{A}^{-1}\right)_{ki}\left(\mathbf{A}^{-1}\right)_{jl}f_l \tag{6.20}$$

so that

$$\frac{\partial s_k}{\partial a_{ij}} = -\left(\mathbf{A}^{-1}\right)_{ki}s_j \tag{6.21}$$

for the coefficients of the vector $\dfrac{\partial \mathbf{s}}{\partial a_{ij}}$. It may be noted that Harville (1997, p.511) also gives a differentiation rule for the pseudoinverse, which is an extended form of that for the normal inverse. However, as we will almost invariably be able to employ the normal inverse, the discussion will here be restricted to the simpler form. It may also be noted that Heijungs (1994) gives a derivation on the basis of Cramer's rule (see Section 4.1.4) that leads to equivalent results.

Knowledge of these derivatives opens the way to explore sensitivity and uncertainty analyses. It may be appropriate to investigate the properties of the derivatives $\dfrac{\partial s_k}{\partial a_{ij}}$ for all values of i and j. We can arrange these in a matrix that we indicate by $\dfrac{\partial s_k}{\partial \mathbf{A}}$, and which is defined as

$$\frac{\partial s_k}{\partial \mathbf{A}} = \begin{pmatrix} \dfrac{\partial s_k}{\partial a_{11}} & \dfrac{\partial s_k}{\partial a_{12}} & \cdots \\ \dfrac{\partial s_k}{\partial a_{21}} & \dfrac{\partial s_k}{\partial a_{22}} & \cdots \\ \cdots & \cdots & \cdots \end{pmatrix} \tag{6.22}$$

We can also extend this form from the scaling vector $\mathbf{s}$ to the inventory vector $\mathbf{g}$. Because

$$\mathbf{g} = \mathbf{B}\mathbf{s} \tag{6.23}$$

the derivatives are given by

$$\frac{\partial \mathbf{g}}{\partial a_{ij}} = \frac{\partial \mathbf{B}}{\partial a_{ij}} \mathbf{s} + \mathbf{B} \frac{\partial \mathbf{s}}{\partial a_{ij}} = \mathbf{B} \frac{\partial \mathbf{s}}{\partial a_{ij}} \tag{6.24}$$

which, by inserting the previously obtained expression for $\dfrac{\partial \mathbf{s}}{\partial a_{ij}}$ gives

$$\frac{\partial g_k}{\partial a_{ij}} = - \sum_l b_{kl} \left(\mathbf{A}^{-1} \right)_{li} s_j \tag{6.25}$$

which can written as

$$\frac{\partial g_k}{\partial a_{ij}} = - \left(\mathbf{B}\mathbf{A}^{-1} \right)_{ki} s_j = -\lambda_{ki} s_j \tag{6.26}$$

Again, it may be convenient to define a matrix $\dfrac{\partial g_k}{\partial \mathbf{A}}$ as

$$\frac{\partial g_k}{\partial \mathbf{A}} = \begin{pmatrix} \dfrac{\partial g_k}{\partial a_{11}} & \dfrac{\partial g_k}{\partial a_{12}} & \cdots \\ \dfrac{\partial g_k}{\partial a_{21}} & \dfrac{\partial g_k}{\partial a_{22}} & \cdots \\ \cdots & \cdots & \cdots \end{pmatrix} \tag{6.27}$$

Possible uses of this expression are discussed in Section 8.2.3.

6.1.2 Perturbation of the intervention matrix

The previous section dealt with perturbations of the technology matrix $\mathbf{A}$. We may also perturb the intervention matrix $\mathbf{B}$. The derivatives are needed in

$$\frac{\partial \mathbf{g}}{\partial b_{ij}} = \frac{\partial \mathbf{B}}{\partial b_{ij}} \mathbf{s} + \mathbf{B} \frac{\partial \mathbf{s}}{\partial b_{ij}} = \frac{\partial \mathbf{B}}{\partial b_{ij}} \mathbf{s} \tag{6.28}$$

This then yields

$$\frac{\partial g_k}{\partial b_{ij}} = \begin{cases} s_j & \text{if } i = k \\ 0 & \text{otherwise} \end{cases} \tag{6.29}$$

Once more, we write this as a matrix $\dfrac{\partial g_k}{\partial \mathbf{B}}$:

$$\frac{\partial g_k}{\partial \mathbf{B}} = \begin{pmatrix} \dfrac{\partial g_k}{\partial b_{11}} & \dfrac{\partial g_k}{\partial b_{12}} & \cdots \\ \dfrac{\partial g_k}{\partial b_{21}} & \dfrac{\partial g_k}{\partial b_{22}} & \cdots \\ \cdots & \cdots & \cdots \end{pmatrix} \tag{6.30}$$

Again, Section 8.2.3 discusses the use of these expressions.

6.1.3 Perturbation of the final demand vector

We may finally start by perturbing the final demand vector $\mathbf{f}$ instead of the technology matrix $\mathbf{A}$ or the intervention matrix $\mathbf{B}$, so that

$$\mathbf{A} \left(\mathbf{s} + \delta \mathbf{s} \right) = \mathbf{f} + \delta \mathbf{f} \tag{6.31}$$

Obviously

$$\mathbf{s} + \delta \mathbf{s} = \mathbf{A}^{-1} \left(\mathbf{f} + \delta \mathbf{f} \right) \tag{6.32}$$

hence

$$\delta \mathbf{s} = \mathbf{A}^{-1} \left(\mathbf{f} + \delta \mathbf{f} \right) - \mathbf{s} = \mathbf{A}^{-1} \delta \mathbf{f} \tag{6.33}$$

In the limit of infinitesimally small $\delta \mathbf{f}$ this reduces to

$$\frac{\partial \mathbf{s}}{\partial f_i} = \mathbf{A}^{-1} \frac{\partial \mathbf{f}}{\partial f_i} \tag{6.34}$$

or

$$\frac{\partial s_k}{\partial f_i} = \left(\mathbf{A}^{-1}\right)_{ki} \tag{6.35}$$

In addition

$$\frac{\partial \mathbf{g}}{\partial f_i} = \mathbf{B}\mathbf{A}^{-1} \frac{\partial \mathbf{f}}{\partial f_i} \tag{6.36}$$

or

$$\frac{\partial g_k}{\partial f_i} = \lambda_{ki} \tag{6.37}$$

These expressions are not of much interest from both a theoretical and a practical aspect. The reason is that the reference flow ϕ and hence the final demand vector $\mathbf{f}$ are often set at a fixed and in fact arbitrary value, such as 1000 kWh of electricity or 100 km of car-driving; there are no uncertainties in the numerical specification. However, these forms are appropriate to introduce an important concept in matrix perturbation theory: the condition number. Because

$$\delta \mathbf{s} = \mathbf{A}^{-1} \delta \mathbf{f} \tag{6.38}$$

the inequality for multiplications of norms yields

$$\|\delta \mathbf{s}\| \leq \|\mathbf{A}^{-1}\| \, \|\delta \mathbf{f}\| \tag{6.39}$$

We also have

$$\|\mathbf{f}\| \leq \|\mathbf{A}\| \, \|\mathbf{s}\| \tag{6.40}$$

These two equations can be combined to give

$$\frac{\|\delta \mathbf{s}\|}{\|\mathbf{s}\|} \leq \|\mathbf{A}\| \, \|\mathbf{A}^{-1}\| \, \frac{\|\delta \mathbf{f}\|}{\|\mathbf{f}\|} \tag{6.41}$$

The factor $\|\mathbf{A}\| \, \|\mathbf{A}^{-1}\|$ thus provides an upper bound of the magnification factor that measures how a relative change in $\mathbf{f}$ propagates as a relative change in $\mathbf{s}$. This factor is known as the condition number of the matrix $\mathbf{A}$, and the symbol $\kappa(\mathbf{A})$ is used to denote it:

$$\kappa(\mathbf{A}) = \|\mathbf{A}\| \, \|\mathbf{A}^{-1}\| \tag{6.42}$$

Note that the condition number has several important characteristics:

- it gives an upper bound for the magnification factor;

- it is defined as the product of two matrix norms, which are themselves defined with the use of a maximum vector norm;

- it is defined as the property of a matrix, and does not consider the right-hand side (here: $\mathbf{f}$) of the system of equations (even though it has been derived under the perturbation of $\mathbf{f}$);

- it is an overall property of a matrix, and it does not indicate which coefficients of $\mathbf{A}$ or $\mathbf{f}$ are most 'sensitive.'

As a consequence, the condition number in certain cases "may tend to grossly overstate" (Sebald, 1974, p.35) the influences of perturbations.

6.1.4   Some examples

Let us return to the old example. The technology matrix is

$$\mathbf{A} = \begin{pmatrix} -2 & 100 \\ 10 & 0 \end{pmatrix} \tag{6.43}$$

Its norm is

$$\|\mathbf{A}\| = 100.02\cdots \tag{6.44}$$

and the norm of its inverse

$$\|\mathbf{A}^{-1}\| = 0.10\cdots \tag{6.45}$$

The product of these two norms is the condition number:

$$\kappa(\mathbf{A}) = 10.004\cdots \tag{6.46}$$

This is not a very remarkable value for a condition number. But, let us consider a different system with the following technology matrix:

$$\mathbf{A} = \begin{pmatrix} -2 & 100 \\ 10 & -499 \end{pmatrix} \tag{6.47}$$

Multiplying its inverse with $\mathbf{f}$ yields

$$\mathbf{s} = \begin{pmatrix} 50000 \\ 1000 \end{pmatrix} \tag{6.48}$$

A small perturbation of the element in the second row and the second column might yield

$$\mathbf{A'} = \begin{pmatrix} -2 & 100 \\ 10 & -498 \end{pmatrix} \tag{6.49}$$

with scaling factors

$$\mathbf{s'} = \begin{pmatrix} 25000 \\ 500 \end{pmatrix} \tag{6.50}$$

Hence, the perturbation of only one of the coefficients by less than 1% propagates as a perturbation of the results by 50%. The condition number of the unperturbed matrix is $130,000$, which points out that this system is likely to be sensitive (or even more extremely) to small perturbations. Let us now study the actual perturbations as indicated by the matrices $\dfrac{\partial s_k}{\partial \mathbf{A}}$. For the first system, we have

$$\frac{\partial s_1}{\partial \mathbf{A}} = \begin{pmatrix} 0 & 0 \\ -10 & -0.2 \end{pmatrix} \tag{6.51}$$

and

$$\frac{\partial s_2}{\partial \mathbf{A}} = \begin{pmatrix} -1 & -0.02 \\ -0.2 & -0.004 \end{pmatrix} \tag{6.52}$$

and for the second system

$$\frac{\partial s_1}{\partial \mathbf{A}} = \begin{pmatrix} -1.25 \times 10^7 & -2.5 \times 10^5 \\ -2.5 \times 10^6 & -5 \times 10^4 \end{pmatrix} \tag{6.53}$$

and

$$\frac{\partial s_2}{\partial \mathbf{A}} = \begin{pmatrix} -2.5 \times 10^5 & -5 \times 10^3 \\ -5 \times 10^4 & -1 \times 10^3 \end{pmatrix} \tag{6.54}$$

Similarly, we have

$$\frac{\partial g_1}{\partial \mathbf{A}} = \begin{pmatrix} -10 & -0.2 \\ -12 & -0.24 \end{pmatrix} \tag{6.55}$$

and

$$\frac{\partial g_2}{\partial \mathbf{A}} = \begin{pmatrix} -2 & -0.04 \\ -1.4 & -0.028 \end{pmatrix} \tag{6.56}$$

and

$$\frac{\partial g_3}{\partial \mathbf{A}} = \begin{pmatrix} 50 & 1 \\ 10 & 0.2 \end{pmatrix} \tag{6.57}$$

for the perturbation of elements of the technology matrix of the first system, and

$$\frac{\partial g_1}{\partial \mathbf{B}} = \begin{pmatrix} 100 & 2 \\ 0 & 0 \\ 0 & 0 \end{pmatrix} \tag{6.58}$$

and

$$\frac{\partial g_2}{\partial \mathbf{B}} = \begin{pmatrix} 0 & 0 \\ 100 & 2 \\ 0 & 0 \end{pmatrix} \tag{6.59}$$

and

$$\frac{\partial g_3}{\partial \mathbf{B}} = \begin{pmatrix} 0 & 0 \\ 0 & 0 \\ 100 & 2 \end{pmatrix} \tag{6.60}$$

for the perturbation of elements of its intervention matrix. For the second system, the matrices have much larger coefficients.

6.2   Uncertainties and their propagation

Thisted (1988, p.101–102) notes that "roughly speaking, if the input data to a linear system are 'good to t decimal places,' then the solution to the linear system may only be good to $t - \log_{10}(\kappa(\mathbf{X}))$ decimal places." For instance, when the input data is correct to 3 decimal places, a condition of number of 100 means that 2 decimal places are lost in the solution to a system of equations, so that the solution is correct to 1 decimal place only. Given the fact that most data for life cycle inventory analysis are quite imprecise, this is a dramatic observation. In most databases, data for an inventory analysis are good to only one or two decimal places. A technology matrix with a condition number of 10 or 100 would deprive the results from any certainty. Uncertainties in data and their propagation in the computations is therefore of theoretical and practical interest.

6.2.1   Uncertainties in data

Let us start by considering one single unit process. Traditionally, it would be described by a process vector

$$\mathbf{p} = \begin{pmatrix} p_1 \\ p_2 \\ \cdots \end{pmatrix} \tag{6.61}$$

Now, we wish to abandon the idea of point estimates, and include the idea that a distribution of values for the elements of this process vector exists. Such a distribution may be interpreted as arising from variability from day to day, variability from place to place, by random fluctuations, because the process is in fact a collection of individual processes with inherent sampling error, or whatever may be the case. Anyhow, all these coefficients will be assumed to be stochastic variables. In principle, one would need to describe the underlying distributions for a proper accounting. For simplicity, we will restrict the discussion by assuming that these distributions are Gaussian. This enables one to describe the distributions with two parameters, for instance, the mean and the standard deviation. A certain coefficient with a point estimate p would then be replaced by $N(p, \sigma)$, where $N(\cdot)$ the represents the Gaussian or normal probability distribution with mean p and standard deviation σ. In the experimental sciences, it is quite normal to write such a stochastic variable as $p \pm \sigma$, although this notation is sometimes used to indicate the range of possible values that the variable may attain or the variable's 95%-confidence interval. Note that, when σ indeed refers to the standard deviation of a Gaussian distribution, the 95%-confidence interval ranges from $p - 1.96\sigma$ to $p + 1.96\sigma$.

In a stochastic system, we might write a process vector $\tilde{\mathbf{p}}$ as

$$\tilde{\mathbf{p}} = \begin{pmatrix} N(p_1, \sigma_1) \\ N(p_2, \sigma_2) \\ \cdots \end{pmatrix} \tag{6.62}$$

or

$$\tilde{\mathbf{p}} = \mathbf{p} + \delta\mathbf{p} \text{ with } \delta\mathbf{p} = \begin{pmatrix} N(0, \sigma_1) \\ N(0, \sigma_2) \\ \cdots \end{pmatrix} \tag{6.63}$$

In principle, every process vector may be defined in this way, and a stochastic process matrix $\tilde{\mathbf{P}}$ may be constructed.

This is a conservative procedure, as it is based on the idea that all uncertainties are independent (Smith *et al.*, 1992). In practice, this is not true. If the fuel input of a car is specified as 10 ± 2 litre and the release of CO_2 as 100 ± 20 kg, it is probable that a high fuel input will be associated with a high CO_2-release and the other way around. For other flows, the correlation may be negative. At any rate, it is likely that not all uncertainties are independent. In general, the degree of dependence between two random variables is measured by their covariance, the covariance between a variable and itself being the variance:

$$\sigma_1^2 = \text{cov}(\tilde{p}_1, \tilde{p}_1) \tag{6.64}$$

Variances and covariances of a number of variables may be arranged in a matrix that is called the covariance matrix (or variance-covariance matrix, or dispersion matrix) :

$$\boldsymbol{\Sigma} = \begin{pmatrix} \sigma_1^2 & \mathrm{cov}(\tilde{p}_2, \tilde{p}_1) & \cdots \\ \mathrm{cov}(\tilde{p}_1, \tilde{p}_2) & \sigma_2^2 & \cdots \\ \cdots & \cdots & \cdots \end{pmatrix} \tag{6.65}$$

Note that $\mathrm{cov}(\tilde{p}_1, \tilde{p}_2) = \mathrm{cov}(\tilde{p}_2, \tilde{p}_1)$, and that $\boldsymbol{\Sigma}$ is therefore a symmetric matrix.

Despite the reduction of information due to this symmetry, an obvious disadvantage of this approach is that a very large matrix is needed to represent all variances and covariances. In practice, one is already glad to have available variances of process data, while covariances are almost never available in an empirical form. Only theoretical considerations, for instance on the basis of mass balances and chemical reaction theory can provide estimates of covariances.

The fact that data in LCA (process data, characterisation factors, etc.) are uncertain to a certain degree has been acknowledged at many places (*cf.* Fava *et al.* (1994), Hoffman *et al.* (1995), Pohl *et al.* (1996), Huijbregts (1998), Maurice *et al.* (2000)). Also, systems for describing, classifying or coding such uncertainties have been used or developed in many texts on LCA. Main trends include:

- descriptions on the basis of categorical data quality descriptors, *e.g.*, Fava et al, (1994), Weidema & Wesnæs (1996), van den Berg *et al.* (1999);

- descriptions on the basis of distribution theory, *e.g.*, Kennedy *et al.* (1996), Hanssen & Asbjørnsen (1996);

- descriptions on the basis of fuzzy set theory, *e.g.*, Chevalier & Le Téno (1996), Roš (1998), Pohl (1999).

Describing uncertainties in data is to a certain sense similar to describing data. A next step, similar to describing how to combine data into an overall result, is describing how to combine uncertainties in data into an overall uncertainty. For the combination of uncertainties on the basis of categorical descriptors and fuzzy set theory, there is not much available. For the combination of uncertainties on the basis of distribution theory, main approaches are:

- approaches based on parametric variation, *e.g.*, Copius Peereboom *et al.* (1998);

- approaches based on analytical expressions for error propagation, *e.g.*, Heijungs (1994), Kennedy *et al.* (1997), Ciroth (2001);

- approaches based on random experiments for error propagation, *e.g.*, Coulon *et al.* (1997), Huijbregts (1998), Ciroth (2001).

As this book is not concerned with data, it also will not discuss uncertainties in data. But, since the emphasis is on the calculations in which data are combined, the combination of uncertainties is of definite interest for this book. In this section, we will restrict the discussion to analytical expressions for error propagation on the basis of perturbation theory. Monte Carlo techniques, as the main approach to random experiments, are discussed in Section 6.4. In addition, it should be mentioned that there exist a limited number of approaches towards other types of data analysis, see, *e.g.*, Le Téno (1999).

6.2.2 Uncertainties in results

The theory of error propagation is a next ingredient for handling uncertainties (Taylor (1982), Morgan & Henrion (1990)). If two random variables x and y are combined, by addition, multiplication, or whatever, to form a new quantity q, the uncertainties in x and y propagate as an uncertainty in q. In general, if

$$q = q(x, y) \tag{6.66}$$

then the variances add, as a first-order approximation, according to

$$\sigma_q^2 = \left(\frac{\partial q}{\partial x}\right)^2 \sigma_x^2 + \left(\frac{\partial q}{\partial y}\right)^2 \sigma_y^2 + 2\frac{\partial q}{\partial x}\frac{\partial q}{\partial y}\mathrm{cov}(x, y) \tag{6.67}$$

When the covariance between different variables is ignored, this reduces to

$$\sigma_q^2 = \left(\frac{\partial q}{\partial x}\right)^2 \sigma_x^2 + \left(\frac{\partial q}{\partial y}\right)^2 \sigma_y^2 \tag{6.68}$$

When the covariance is not ignored, one may use the relation

$$\mathrm{cov}(x, y) \le \sigma_x \sigma_y \tag{6.69}$$

and find that

$$\sigma_q^2 \le \left(\left|\frac{\partial q}{\partial x}\right| \sigma_x + \left|\frac{\partial q}{\partial y}\right| \sigma_y\right)^2 \tag{6.70}$$

which gives

$$\delta_q \leq \left|\frac{\partial q}{\partial x}\right| \delta_x + \left|\frac{\partial q}{\partial y}\right| \delta_y \tag{6.71}$$

as an upper limit (Taylor (1982)).

Application in LCA can be based on these general results. For instance, as

$$\frac{\partial g_k}{\partial a_{ij}} = -\lambda_{ki} s_j \tag{6.72}$$

it follows that, approximately and ignoring covariances,

$$\sigma^2(g_k) = \sum_{i,j} \lambda_{ki}^2 s_j^2 \sigma^2(a_{ij}) \tag{6.73}$$

In this way, all equations of LCA may be processed.

Another approach is presented by Stewart (1990), who works out the statistical properties of random matrices, or matrices with a stochastic perturbation. The approach allows for a random matrix $\delta\mathbf{A}$ added to a matrix $\mathbf{A}$ with fixed coefficients. In principle, one would like to study the effects of a random matrix $\delta\mathbf{A}$ with arbitrary probability distributions and moments. For instance, one would like to draw $\delta\mathbf{A}_{11}$ from a Gaussian distribution with variance 5 and draw $\delta\mathbf{A}_{12}$ from a uniform distribution with width 13. However, as "random matrices are difficult to manipulate in this generality" (Stewart, 1990, p.582), the discussion is restricted to so-called cross-correlated matrices, defined by

$$\delta\mathbf{A} = \mathbf{c}\mathbf{H}\mathbf{r}^{\mathrm{T}} \tag{6.74}$$

Here $\mathbf{H}$ is a random matrix with elements that are uncorrelated, have a mean 0, and have a variance 1, and the two vectors $\mathbf{c}$ and $\mathbf{r}$ are used to scale the columns and rows respectively of the random elements. Thus, for a specific element $\delta\mathbf{A}_{ij}$ is drawn from a distribution with mean 0 and variance $c_i r_j$. The idea is that the vector $\mathbf{c}$ is flow-specific and that the vector $\mathbf{r}$ is process-specific. The vectors $\mathbf{c}$ and $\mathbf{r}$ measure differences in scales of measurement and differences in precision. For instance, a flow of dioxins in kg will have a smaller uncertainty than a flow of carbon dioxide in kg. Likewise, the process of aluminium production will have a smaller uncertainty than the process of painting a house. Observe that the approach makes no assumptions with respect to the probability distributions. The only assumptions are that elements of $\mathbf{H}$ are themselves uncorrelated. The approach does not use or calculate full distributions, but only the variance of such distributions.

The assumed property of cross-correlation of $\delta\mathbf{A}$ seems workable, at least as a first approximation. Stewart (1990) now proofs that a first-order approximation for $(\mathbf{A} + \delta\mathbf{A})^{-1}$ yields the following:

$$E\left(\left\|\mathbf{s}' - \mathbf{s}\right\|_{\mathrm{F}}^2\right) = \left\|\delta\mathbf{A}\mathbf{c}\right\| \left\|\mathbf{rf}\right\| \tag{6.75}$$

Here $E(\cdot)$ indicates the expectation operator, and $\|\mathbf{X}\|_{\mathrm{F}}$ indicates the Frobenius norm of matrix $\mathbf{X}$. Although this is an interesting result, its practical use for LCA is limited. The reason is that the elements of $\mathbf{s}$ are very different in order of magnitude. Hence, an approximate expression for the norm of $\delta\mathbf{s}$ is less useful than one might be inclined to think.

Still another way is provided by Monte Carlo simulations; for this see Section 6.4.

6.3   Discrete choices

Section 6.2 deals with uncertainties in continuous data. Data are assumed to be uncertain due to variability, measurement error, sampling error, and so on. Data are, however, not always representing a continuous variable. A box of matches contains an integer number of matches. Any modeling by means of continuous distributions is an approximation. A stronger error due to finite granularity occurs when we specify the number of persons in a car. It is 1, 2, 3, 4 or perhaps a few more. Here, one may wish to apply a different type of statistical analysis, in which there is a probability for 1 passenger, a probability for 2 passengers, etc.

In LCA, we have to do with data and assumptions, although the boundary between the two is not always clear. The quantity of carbon dioxide released by a production process is a data item. The choice for a particular technology specification (*e.g.*, coal-fired instead of gas-fired plants) is an assumption. This type of assumption creates uncertainties as well. But they cannot be specified in terms of Gaussian distributions. One way to deal with them is to specify the two technologies along with the probability that this is the appropriate technology. This then is not a probability in the frequentist sense, but more a probability in the personalist sense (*cf.* Gillies (2000)). Probability measures are even more difficult to obtain for this type of uncertainty than for data uncertainty.

Another example of a discrete choice relates to allocation of multifunctional processes. Given one allocation principle, say partitioning on an economic basis, there is data uncertainty because prices may be variable. But the choice between partitioning on an economic basis and on the basis of energy-content is a discrete one. And the choice between partitioning and substitution is likewise discrete.

Morgan & Henrion (1990) discuss the use of scenario trees with qualitative levels to study the propagation of uncertainties due to discrete uncertainties or choices. However, they quickly step to Monte Carlo simulations, because it is easier applied in the computer era than an analytic method, and it may provide more insightful results. Monte Carlo experiments are therefore a logical topic for the next section.

6.4   Monte Carlo simulations

Monte Carlo analysis (see Hendry (1984) for an advanced treatment and Morgan & Henrion (1990) for a more basic treatment) is a sampling technique. That means that a sample of model results is generated, and that statistical properties of this sample, like the mean and the standard deviation, are used to provide an indication of the location and dispersion of the results. Whereas Sections 6.2 and 6.3 put an emphasis on mathematical analysis, Monte Carlo methods relies on numerical simulations. While analytical methods are exact (although approximations are often needed, for instance the restriction to first-order perturbations), sampling techniques provide stochastic results, that are not exactly reproducible. On the other hand, sampling techniques are able to deal with more than only a few theoretical distributions, and can include all types of complicated dependencies. Especially when fast computers are available, Monte Carlo techniques provide a useful means of assessing robustness. LCA is an area where it may be applied fruitfully.

Having specified probability distributions for process data, characterisation factors, normalisation factors, weighting factors, allocation factors, etc. one can proceed to generate one model realisation. This yields an inventory vector, $\mathbf{g}$, that we will denote as $\mathbf{g}^1$. A second realisation yields a second result, $\mathbf{g}^2$. All together, after model realisations, a set of results $\{\mathbf{g}^1, \mathbf{g}^2, \ldots, \mathbf{g}^N\}$ will be obtained. This set of results can be subject to analysis; see Section 8.2.4.

In performing a Monte Carlo analysis, one must choose the number of runs N. It is important to understand that the number of uncertain input parameters does not affect the number of runs required. The number of runs N is only determined by the required accuracy of the output distribution. Morgan & Henrion (1990, p.200 *ff.*) provide useful guidelines for choosing N. As a rule of thumb, they state that "in most uncertainty analysis of quantitative policy models, a few hundred or sometimes only a few tens of runs may be quite sufficient."

In addition to Monte Carlo simulations, alternative sampling methods exist, such as Latin hypercube sampling, in which the sampling space is subdivided into strata. We see no particular advantage in employing these techniques in the case LCA.

6.5 Change of technology

Most of the results of this chapter was based upon perturbations that were small enough to allow for a first-order approximations, so that differential calculus could be employed. Equations such as

$$\frac{\partial \mathbf{s}}{\partial a_{ij}} = -\mathbf{A}^{-1}\frac{\partial \mathbf{A}}{\partial a_{ij}}\mathbf{A}^{-1}\mathbf{f} \tag{6.76}$$

were used at many places. However, it was also noticed that a more accurate expression was available:

$$\delta \mathbf{s} = -\left(\mathbf{A} + \delta \mathbf{A}\right)^{-1}\delta \mathbf{A}\mathbf{A}^{-1}\mathbf{f} \tag{6.77}$$

For changes in the intervention matrix, the form

$$\delta \mathbf{g} = \delta \mathbf{B}\mathbf{s} \tag{6.78}$$

holds. These expressions may be of specific interest when there are changes in technology, *e.g.*, when a cleaner or more efficient process is introduced. This leads to a replacement of a number of coefficients in one column of $\mathbf{A}$ and/or $\mathbf{B}$. This change could be expressed as $\delta \mathbf{A}$ and $\delta \mathbf{B}$. In principle, the above two expressions can be used to calculate new results. In practice, the use of these expressions may be harder to use than straightforward calculation of

$$\mathbf{s} + \delta \mathbf{s} = -\left(\mathbf{A} + \delta \mathbf{A}\right)^{-1}\mathbf{f} \tag{6.79}$$

and

$$\mathbf{g} + \delta \mathbf{g} = \left(\mathbf{B} + \delta \mathbf{B}\right)\mathbf{s} \tag{6.80}$$

However, the use of the two expressions may provide insight on coefficients for which the results are extremely sensitive. This takes place in the perturbation analysis (Section 8.2.3).

6.6   Numerical stability

This section will explore some of the consequences of perturbation-theoretic concepts for the numerical stability of the calculations that are made in the inventory analysis. The main difference with Section 6.2 is that the emphasis is on the uncertainty introduced by computational errors instead of by measurement errors.

6.6.1   Round-off

Let us consider the system of Section 4.3:

$$\mathbf{A} = \begin{pmatrix} -2 & 100 \\ 10 & -10 \end{pmatrix} \tag{6.81}$$

We can compute

$$\mathbf{s} = \mathbf{A}^{-1}\mathbf{f} \tag{6.82}$$

Recall from Section 3.1 the definition of the of final supply vector:

$$\tilde{\mathbf{f}} = \mathbf{A}\mathbf{s} \tag{6.83}$$

and that of the discrepancy vector:

$$\mathbf{d} = \tilde{\mathbf{f}} - \mathbf{f} \tag{6.84}$$

As there is no cut-off, allocation, closed-loop recycling or other special situation, we would expect to find

$$\mathbf{d} = \mathbf{A}\left(\mathbf{A}^{-1}\mathbf{f}\right) - \mathbf{f} = \mathbf{0} \tag{6.85}$$

However, if we type this formula and the appropriate coefficients into Matlab (a computer package that is especially useful for matrix computations), we find

$$\mathbf{d} = \begin{pmatrix} -0.5684 \times 10^{-13} \\ 0 \end{pmatrix} \tag{6.86}$$

This non-zero discrepancy vector arises from computational round-off. In this case, it is so small that we can safely ignore it, but there might be situations in which errors are introduced that are of importance. A study of the effects of using a finite precision, in relation to the way numbers are stored and treated in a computer is therefore of interest. Part of the material related to numerical stability has to do with the choice of algorithm.

This will be discussed in Section 10.2. But another part has to do with perturbation theory. This is treated in this section.

In Section 6.2 it was stated that the logarithm of the condition number provides a rough measure of the number of significant digits that are lost in a matrix inversion. The matrix in the example above has a condition number of 10.3. If a computer system works with an internal representation of 7 or 10 digits, we will not loose precision in the calculations. Suppose, however, that we measure the first flow for some reason not as litre of fuel, but instead in 10^{-11} litre of fuel. The technology matrix is then

$$\mathbf{A} = \begin{pmatrix} -2 \times 10^{-11} & 100 \times 10^{-11} \\ 10 & -10 \end{pmatrix} \tag{6.87}$$

it has a condition number of 10^{12}, and the discrepancy vector becomes

$$\mathbf{d} = \begin{pmatrix} -0.0039 \\ 0 \end{pmatrix} \tag{6.88}$$

With a large condition number, we may loose precision which does matter.

6.6.2 Rescaling

From the previous example, we see that a change of scale affects the condition number and hence the round-off danger in numerical operations. We may use this to our advantage, as a rescaling of rows and/or columns is always possible if carried out consistently. This proceeds by

$$a'_{ij} = r_i a_{ij} c_j \tag{6.89}$$

where r_i represents the factor with which row i is scaled, and c_j the factor with which column j is scaled. This can also be written as

$$\mathbf{A}' = \mathrm{diag}(\mathbf{r})\mathbf{A}\mathrm{diag}(\mathbf{c}) \tag{6.90}$$

The rules to be applied to $\mathbf{s}$ and $\mathbf{f}$ follows directly from

$$\mathbf{A}'\mathbf{s}' = \mathbf{f}' \Leftrightarrow \mathrm{diag}(\mathbf{r})\mathbf{A}\mathrm{diag}(\mathbf{c})\,(\mathrm{diag}(\mathbf{c}))^{-1}\mathbf{s} = \mathrm{diag}(\mathbf{r})\mathbf{f} \tag{6.91}$$

In other words,

$$\mathbf{f}' = \mathrm{diag}(\mathbf{r})\mathbf{f} \tag{6.92}$$

and

$$\mathbf{s}' = (\mathrm{diag}(\mathbf{c}))^{-1}\mathbf{s} \tag{6.93}$$

In some cases, one may be more interested in the inverse matrix $\mathbf{A}^{-1}$ than in the scaling vector $\mathbf{s}$. Transformation rules for rescaling the inverse matrix are obtained from the fact that

$$(\mathbf{diag}(\mathbf{r})\mathbf{A}\,\mathbf{diag}(\mathbf{c}))^{-1} = (\mathbf{diag}(\mathbf{c}))^{-1}\,\mathbf{A}^{-1}\,(\mathbf{diag}(\mathbf{r}))^{-1} \tag{6.94}$$

from which it follows that

$$\begin{aligned} \mathbf{A}^{-1} &= \mathbf{diag}(\mathbf{c})\,(\mathbf{diag}(\mathbf{r})\mathbf{A}\,\mathbf{diag}(\mathbf{c}))^{-1}\,\mathbf{diag}(\mathbf{r}) \\ &= \mathbf{diag}(\mathbf{c})\mathbf{A}'^{-1}\mathbf{diag}(\mathbf{r}) \end{aligned} \tag{6.95}$$

In the example system, the choice

$$\mathbf{r} = \begin{pmatrix} 0.1 \\ 0.01 \end{pmatrix} \text{ and } \mathbf{c} = \begin{pmatrix} 1 \\ 1 \end{pmatrix} \tag{6.96}$$

yields a substantial reduction of the condition number from $\kappa(\mathbf{A}) = 10$ to $\kappa(\mathbf{A}') = 1.2$.

As a rule of thumb, one may say that $\mathbf{r}$ and $\mathbf{c}$ must be chosen such that the absolute value most elements of $\mathbf{A}$ lies between 0.01 and 100. Especially elements of $\mathbf{r}$ may well be chosen as 10^{-9}, 10^{-6}, 0.001, 1000, 10^6, 10^9, etc. with an interpretation of changing the units for the rows from kg into mg, mg, g, Mg, Gg, Tg, and so on. Another option for rescaling is to standardise every column so that its mean value is zero and its variance one (see Neter et al. (1996, p.277 ff.)). The problem of optimal rescaling is unresolved; see Golub & Van Loan (1996, p.125) and Forsythe & Moler (1967, p.37 ff.) for some useful discussions.

Chapter 7

Structural theory

A technology matrix can be said to be a compilation of process data, a database. One must keep in mind, however, that the processes are linked to each other. It is not a database of processes, but a database of linked processes. The linkage of the processes is expressed in terms of economic flows, produced by one process, absorbed by a different process. A technology matrix therefore contains more than process data: it also contains information on the structure of the interindustry dependence of processes. This is also true for input-output analysis (see Chapter 5): Leontief's grand work bears the title *The structure of American economy*. This chapter will discuss approaches towards exploring the structure of a life cycle or product system. The background of such explorations is provided by something that we will call structural theory. It is a topic that has hardly been addressed in the context of LCA. Huele & Van den Berg (1998) and Le Téno (1999) probably present the only approaches in which the analysis of the structure of an LCA-database is addressed. Structural theory can be said to be the least-developed topic. This chapter can therefore not provide a state-of-the-art review. We can do no better than discuss a number of approaches that have been described, that can be borrowed from related fields of study (like input-output analysis), or that we believe to be promising.

7.1 Some definitions

It is natural to use the technology matrix, its inverse and/or the intervention matrix in carrying out analyses as to the structure of a product system. There are various candidates for this. We start by reviewing these.

The technology matrix $\mathbf{A}$ is the purest representation of the structure of the structure of the economy in terms of unit processes. Especially when the processes are recorded on the basis of flows per year instead of flows per unit of output (see also Section 2.6) does the technology matrix provide a picture of the volume of activity of every process in relation to the other processes.

Chapter 3 provides procedures to cut off flows for which no production data are available and to deal with multifunctional processes. This involves several manipulations to transform a technology matrix $\mathbf{A}$ into a matrix $\mathbf{A}'$ that will be referred to as the allocated technology matrix (although one should keep in mind that the manipulations may include more than allocation only).

The inventory problem is in general $\mathbf{A}'\mathbf{s} = \mathbf{f}$. For one particular scaling economic flow, this may be written as

$$\sum_j a'_{ij}s_j = f_i \tag{7.1}$$

Once the scaling vector $\mathbf{s}$ is known, we may single out the individual terms $a'_{ij}s_j$ of this summation and construct

$$\mathbf{As} = \begin{pmatrix} a'_{11}s_1 & a'_{12}s_2 & \cdots \\ a'_{21}s_1 & a'_{22}s_2 & \cdots \\ \cdots & \cdots & \cdots \end{pmatrix} \tag{7.2}$$

This matrix can also be written as $\mathbf{A}'\mathbf{diag}(\mathbf{s})$. It contains all columns of the (allocated) technology matrix scaled with the appropriate scaling factor, and it will therefore be referred to as the scaled technology matrix.

The solution of the inventory problem itself involves the inverse of the allocated technology matrix, $\mathbf{A}'^{-1}$.

As for the environmental flows, we may, like for the technology matrix, distinguish the pure, unallocated intervention matrix $\mathbf{B}$, the allocated intervention matrix $\mathbf{B}'$, and the scaled intervention matrix $\mathbf{B_S} = \mathbf{B}'\mathbf{diag}(\mathbf{s})$ which displays each flow in the appropriate magnitude.

There is no such thing as the inverse of the intervention matrix, Of interest, however, may be the product of the allocated intervention matrix and the inverse of the allocated technology matrix. We have defined this matrix in Section 2.2 under the name intensity matrix $\mathbf{\Lambda}$.

7.2   Summary measures

Summary measures are numbers that condense a certain aspect of a large set of numbers into one or perhaps a few items. Many statistics are in fact summary numbers: the mean and the standard deviation of a data set provide two characteristic features that may be interpreted as a measure of central tendency and a measure of dispersion respectively. Such statistics have been discussed in the context of a Monte Carlo analysis; see Section 8.2.4. They cannot serve to provide a measure of the structure of a system. For this, several other summary measures may be considered. Below, we discuss some of these.

7.2.1 The condition number

In Section 6.1.3 the condition number of the (allocated) technology matrix was introduced as a number that provides an upper bound of the overall sensitivity of the system to perturbation of the coefficients. Although its definition has been formulated in the context of perturbation theory, the condition number is a property of a matrix that can be calculated without the need to perturb coefficients. It is an overall property of a matrix, and its value depends, besides on the scale of the rows and columns, of the structure of the dependencies between the rows and columns. It can therefore also be regarded as a summary measure of the structure of a technology matrix. The interpretation of the condition number as an indicator of the structure of a system is, however, not clear.

7.2.2 Other characteristics

In addition to the condition number, there are additional characteristic properties of a matrix. We mention just a few:

- the determinant;
- the trace;
- the rank;
- the spectral radius.

We will not exclude the possibility that one of these properties for one of the matrices mentioned in Section 7.1 may provide useful information that may be given an interpretation of the structure of the system under study. But, as a matter of fact, we have not yet been able to devise such interpretations. Of course, the rank is an indication of the degree of independence, and the spectral radius has been discussed in Section 4.2 as providing an indication whether a power series expansion is possible. But, up to now, it remains unclear how to relate this information to the system's structure.

7.2.3   Correlation analysis

Pearson's product moment correlation coefficient, or the correlation coefficient in short, of two data sets of equal length is a dimensionless number
that lies between -1 and 1. It measure the degree of linear association
between two variables. Perfect association yields a correlation coefficient 1
in case of a positive slope and -1 in case of a negative slope. A correlation coefficient 0 indicates totally uncorrelated variables. An operational
formula for the correlation coefficient of two series x and y is

$$r = \frac{\sum_{i=1}^{N}(x_i - \bar{x})(y_i - \bar{y})}{\sum_{i=1}^{N}(x_i - \bar{x})^2 \sum_{i=1}^{N}(y_i - \bar{y})^2} \tag{7.3}$$

Correlation analysis has been applied by Huele & Van den Berg (1998).
Their object of analysis is the matrix that has been indicated as $\mathbf{G_I}$ in
section 3.8.2, which has been shown to be equal to the intensity matrix
$\mathbf{\Lambda}$. They direct attention to those pairs of interventions that are perfectly
and positively correlated, *i.e.* for which $r = 1$. Possible reasons for this
correlation are:

- the correlation coefficient is not really 1 but has been rounded towards
  1;

- some interventions appear only in one process.

In addition, we should mention the following:

- the possibility that errors in nomenclature have been made;

- the fact that generic factors have been used, *e.g.*, in converting emissions of NO_x into NO, NO_2, etc.

Huele & Van den Berg (1998) also study those pairs of inventory tables
for which $r = 1$. This points out those inventory tables that have been
assumed to be the same, for lack of more specific data, like electricity
generation in Belgium, Spain and Greece.

7.2.4   Multivariate methods

Heijungs *et al.* (1992, p.II-118) ask themselves, in the context of contribution analysis, the following: "can the search for major axes (= dominant

processes?) be made easier by singular value decomposition?" Huele & Van den Berg (1998, p.114) mention that "a simple factor analysis was carried out," but their paper does not provide more clues than a future outlook (p.118). Heijungs & Kleijn (2001, p.148) also "mention the enormous toolbox of multivariate methods, like principal component analysis, factor analysis and cluster analysis" as an item of interest, but do not provide more details. Finally, Le Téno (1999) reveals a paper in which principal component analysis is applied. It appears that there is 'something in the air' in using multivariate methods in analysing and interpreting the structure of LCA. Here, we will describe more, but admit that experience in applying such methods in the context of LCA is lacking almost entirely.

Multivariate analysis (see, *e.g.*, Howard (1991), Johnson & Wichern (1992) and Legendre & Legendre (1998)) represents a class of analytical techniques that aim to detect structural patterns in a data set that can be arranged in a rectangular matrix, say $\mathbf{X}$ of dimension $m \times n$. The structure of this so-called data matrix is thus

$$\mathbf{X} = \begin{pmatrix} x_{11} & x_{12} & \cdots & x_{1n} \\ x_{21} & x_{22} & \cdots & x_{2n} \\ \cdots & \cdots & \cdots & \cdots \\ x_{m1} & x_{m2} & \cdots & x_{mn} \end{pmatrix} \tag{7.4}$$

Multivariate methods are often, but not exclusively, employed in the context of statistical models. The interpretation of the matrix is then that there are m observations on n attributes or variates. Often, but again not exclusively, $m > n$, *i.e.* there are more observations than variates. Although some of the variates may in principle be measured on a categorical (*i.e.* nominal or ordinal) scale, we will restrict the discussion here to the case of strictly cardinal variables. Furthermore, many procedures require a transformation of the data matrix: each column vector of observations is centred, *i.e.* $\mathbf{x}_1$ is replaced by $\mathbf{x}_1 - \bar{\mathbf{x}}_1$.

Principal component analysis (PCA) is an analytical tool that effectively rotates the basis vectors that span a linear space in such a way that the data points are 'better' (in a well-defined sense) described in the linear space that is spanned by the new basis vectors. The general equation for PCA is

$$\mathbf{X}' = \mathbf{X}\mathbf{R} \tag{7.5}$$

where $\mathbf{X}'$ is the data matrix in the rotated system, and $\mathbf{R}$ is an orthogonal matrix that provides an 'optimal' rotation. This rotation is chosen such as to maximise the variances of the data matrix and to minimise the covariances. That is, the covariance matrix of the original data matrix

$$\Sigma = \frac{1}{n-1}\mathbf{X}^T\mathbf{X} \tag{7.6}$$

is transformed into

$$\Sigma' = \frac{1}{n-1}\mathbf{X}'^T\mathbf{X}' = \mathbf{R}^T\Sigma\mathbf{R} \tag{7.7}$$

with zero off-diagonals. The theory of PCA provides expressions for $\mathbf{R}$ and Σ'. $\mathbf{R}$ is the matrix of eigenvectors of $\mathbf{X}$ and the diagonal elements of Σ' are the associated eigenvalues.

The essential idea of PCA may be grasped from a simple example. Suppose that we have a two-dimensional linear space of environmental flows, spanned by

$$\begin{pmatrix} \text{kg of carbon dioxide} \\ \text{kg of sulphur dioxide} \end{pmatrix} \tag{7.8}$$

Also suppose that a number of product systems have been subject to an inventory analysis, and that the results are known as inventory vectors $\mathbf{g}_1, \mathbf{g}_2, \ldots$. Then we may visualise each of these inventory vectors as points in the two-dimensional environmental space. Let us assume that electricity and transport are two of the economic flows for which the inventory vector has been calculated. It is quite probable that all other inventory vectors are to a large extent determined by electricity and transport, not by the processes of electricity production and transportation, but by the system-wide flows related to electricity and transport. Some of the inventory vectors will be dominated by electricity, others by transport, while a third group may have a large contribution of both. A principal component analysis will now enable one to determine the number of dominating constituent life cycles.

As an example, consider the matrix

$$\mathbf{G_I} = \boldsymbol{\Lambda} = \begin{pmatrix} 10 & 0.2 & 12 & 4.6 \\ 2 & 0.1 & 1.8 & 2 \end{pmatrix} \tag{7.9}$$

A first thing to note is that the matrix $\boldsymbol{\Lambda}$ does not have the structure required by most multivariate models. Each inventory vector can be seen as an observation in a space defined by environmental flows. Thus, we have to interchange rows and columns of $\boldsymbol{\Lambda}$, i.e. to transpose it. Indeed, the matrix

$$\boldsymbol{\Lambda}^T = \begin{pmatrix} 10 & 2 \\ 0.2 & 0.1 \\ 12 & 1.8 \\ 4.6 & 2 \end{pmatrix} \tag{7.10}$$

does possess the required structure. The covariance matrix is

$$\Sigma = \begin{pmatrix} 28.5 & 3.76 \\ 3.76 & 0.85 \end{pmatrix} \tag{7.11}$$

A principal components analysis yields the largest eigenvalue $\lambda_1 = 29$ with associated eigenvector $\mathbf{r}_1 = \begin{pmatrix} 0.99 & 0.13 \end{pmatrix}^{\mathrm{T}}$ and second eigenvalue $\lambda_2 = 0.35$ with associated eigenvector $\mathbf{r}_2 = \begin{pmatrix} 0.13 & -0.99 \end{pmatrix}^{\mathrm{T}}$.

We may interpret this as follows. Almost the entire variation in the distribution of product alternatives in the 2-dimensional space can be accounted for by the first principal axis, which is a combination of 0.99 kg of carbon dioxide and 0.13 kg of sulphur dioxide. This axis has no meaning in itself, but it can be given a meaning, by relating it to one of the inventory vectors that falls almost exactly on this axis. Assuming that this is the inventory vector for electricity, we find that electricity is an extremely dominant aspect in many inventory vectors. Figure 7.1 provides a visual interpretation of the aim and results of principal components analysis in the context of LCA.

For a multi-dimensional space, things work out more complicated, but not essentially different. In each principal component analysis, but in particular in those in a high dimensional space with a large number of high eigenvalues, the interpretation of the first few major axes can become difficult.

Factor analysis, also referred to as common factor analysis (CFA), is a technique that is often presented as a further refinement of principal component analysis in a purely statistical context. Because PCA itself is not yet quite understood in relation to LCA, we will not discuss CFA at this stage of development. It appears logical, however, to explore the use of CFA and related methods of multivariate analysis in near future.

7.3 Visual tools

Summary measures as discussed above may sometimes be too crude in reducing a system with a few hundreds of processes and an equally large number of economic and/or environmental flows to one number. For instance, in the context on perturbation theory it was noted that the condition number often overestimates the sensitivity of a system, and moreover does not indicate which data items are the most sensitive ones. Matrices such as $\frac{\partial g_k}{\partial \mathbf{A}}$ provide much more valuable information. However, these matrices are difficult to oversee for a large system. Visual tools are perhaps indispensable in such cases. They do not provide accurate numerical results, but they allow for a quick overview of the most important features.

7.3.1   Spy plots

Huele & Van den Berg (1998) introduce "spy plots" (the name probably derives from Matlab's spy command for viewing sparsity patterns in matrices) as a visual tool to inspect the structure of an database of inventory tables. Their object of analysis is again the intensity matrix Λ. Their Figure 1 can be seen as the transpose of the intensity matrix, with a line denoting a row or column that contains zeros only. We transpose their figure to be able to tie the discussions to the matrix structure in a straightforward manner. Rows of Λ with zeros everywhere indicate "emissions/resources that do not appear in any inventory," while columns with zeros everywhere indicate "inventories that contain no emissions/resources." Empty rows thus point out interventions that either could have been omitted from the system, or that may have been erroneously forgotten. Empty columns mostly refer to partial LCAs with absent data, for instance gate-to-grave waste treatment of used products for which no data is available. Although Huele & Van den Berg (p.117) see spy plots as "a useful way to visualise the presence of empty inventories and non used emissions/resources in an LCA data base," we think that the usefulness is restricted to error checking in practice, and that it cannot be regarded as a genuine tool for structural analysis.

7.3.2   Other graphics

Section 8.2 discusses various ways of interpretation, and Heijungs & Kleijn (2001) propose various sorts of graphical presentation for contribution analysis, uncertainty analysis and comparative analysis. Although such methods for presentation are valuable in themselves in a concrete case study, they cannot be regarded as forms of structural analysis.

A similar argument applies to the visual inspection of disparateness of product alternatives, as in the principal component analysis with bootstrap by Le Téno (1999). The question of disparateness of product alternatives is a crucial one for decision-analysis. We think, however, that the degree of discernibility can better be answered with a statistical analysis than a visual one. Section 8.2.7 introduces such a statistical analysis under the name of discernibility analysis.

Le Téno (1999) also suggests to plot various alternatives in a linear space of which the basis is spanned by environmental flows. This is an interesting addition as a visual aid in interpreting the structure of an intensity matrix with principal component analysis; see Figure 7.1. As long as the rotated

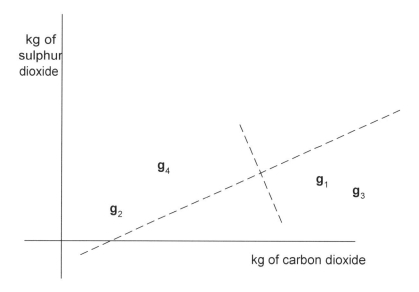

Figure 7.1: Impression of the position of four inventory vectors in a two-dimensional linear space representing two environmental flows. The dashed lines represent the rotated co-ordinate system after principal component analysis.

space is restricted to two dimensions, corresponding to the two eigenvectors with the largest eigenvalues, this works fine. When two directions do not suffice to account for most of the variation, the restrictions of using a planar projection will make visual interpretation cumbersome. We don't know to which extent this will be the case in real-world situations.

Chapter 8

Beyond the inventory analysis

In this chapter, we discuss the computational aspects of the phases and elements of the ISO-framework for LCA which may take place after the inventory analysis. In principle, this concerns the phases impact assessment and interpretation. However, by treating the statistical analysis of the inventory analysis in the previous chapter, as an integral aspect of the inventory analysis, a substantial part of the theoretical part of the interpretation has already been discussed. Moreover, a large number of issues for interpretation deal with procedures, quality checks or reiterations of previous steps. Those issues obviously fall outside the scope of this book. This chapter thus deals with the impact assessment, including its statistical analysis, and selected elements of the interpretation.

8.1 Impact assessment

The theoretical literature on the methodology of LCA is much larger for impact assessment than it is for inventory analysis. Hauschild & Wenzel (1998), for instance, devote 540 pages to impact assessment, and 25 pages to inventory analysis. Pertinent problems arise in impact assessment: which impact categories should be chosen, which category indicators provide an environmentally relevant, yet feasible, representation of environmental impact, which level of spatial and temporal differentiation should be aimed at, how should subjective preferences be incorporated, etc. The computational structure of impact assessment, however, is much easier than that of the inventory analysis. The inventory analysis can be seen as a model, which includes all types of model complications: cut-off, multifunctionality, and so on. Impact assessment, in contrast, is a much more mechanical procedure that uses the results of all sorts of complicated models, provided that such models are available. In other words, the model content of impact assessment is much lower, because it is assumed that the characterisation models that are used in impact assessment cover such modeling technicalities.

In this section, we discuss the computational structure of life cycle impact assessment. This means that certain important steps from the ISO 14042-standard (ISO, 2000) are not or only superficially discussed. Emphasis is on the steps in which calculations are made, *viz.* characterisation, normalisation and weighting. In contrast to inventory analysis, the computational steps of impact assessment are presented in many texts. For instance, Lindfors *et al.* (1995), Heijungs & Hofstetter (1996) and Wenzel *et al.* (1998) offer formulas for characterisation of the form

$$h_i = \sum_j q_{ij} g_j \tag{8.1}$$

for normalisation of the form

$$\tilde{h}_i = \frac{h_i}{\dot{h}_i} \tag{8.2}$$

and for weighting of the form

$$W = \sum_i w_i h_i \tag{8.3}$$

These three formulas are derived in the Sections 8.1.4, 8.1.5 and 8.1.7. However, prior to these steps, some formal aspects of impact categories in relation to linear spaces and the mathematics of characterisation factors should be pointed out.

8.1.1 Impact categories and characterisation models

In a methodological sense, the selection of impact categories is controversial. From a mathematical perspective, however, it is easy. We can just assume that P categories have been chosen, and that associated with these impact categories P category indicators have been chosen. We can then set up a linear space with a P-dimensional basis. For instance, when $P = 3$ and the impact categories are acidification, global warming and resource depletion, with associated category indicators H^+-release, infrared absorption and decreased availability, and units kg SO_2-equivalent, kg CO_2-equivalent and RDU (= resource depletion units, a hypothetical unit for aggregating resources), the basis is

$$\begin{pmatrix} \text{kg } SO_2\text{-equivalent of } H^+\text{-release} \\ \text{kg } CO_2\text{-equivalent of infrared absorption} \\ \text{RDU of reduced availability} \end{pmatrix} \tag{8.4}$$

The indicator results are the numerical outcomes for a certain product alternative. How to find these is described in Section 8.1.4. Here we limit ourselves to the fact that a vector of indicator results defines a point in this linear space. The symbol $\mathbf{h}$ will be used to indicate such a vector. For instance, the vector

$$\mathbf{h} = \begin{pmatrix} 5 \\ 20 \\ 2 \end{pmatrix} \tag{8.5}$$

indicates a vector of indicator results that corresponds to 5 kg SO_2-equivalent of H^+-release, 20 kg CO_2-equivalent of infrared absorption and 2 RDU of reduced availability.

Along with the choice of a category indicator that is supposed to reflect a certain impact category, a characterisation model is taken from literature or developed. A characterisation model is a function that maps a point in $\mathcal{G}$ space to a point in $\mathcal{H}$ space:

$$\eta : \mathcal{G} \rightarrow \mathcal{H} \tag{8.6}$$

We will restrict the discussion to the case that such a point lies on one of the axes of the linear space. In that case, its co-ordinates are 0 for all but one of the vector elements. Thus, given a set of environmental interventions, a characterisation model returns a number, h_i. We will write

$$h_i = \eta_i(\mathbf{g}) \tag{8.7}$$

where η_i is the characterisation model for category indicator (and hence impact category) i.

8.1.2 Derivation of characterisation factors*

In general, η_i will be a non-linear function of its argument vector $\mathbf{g}$. This means that the complete function (and hence the complete model) must be available when an LCA is carried out. This is very inconvenient in practice.

Therefore, a simplification is introduced. It is based on the assumption that there is a relatively large background of environmental interventions, and that one product alternative creates only a small change. It seems therefore reasonable to use a Taylor's series expansion of the function η_i around the background level. This background intervention level will be denoted by $\mathbf{g}_0$. We thus study

$$
\begin{aligned}
\eta_i(\mathbf{g}_0 + \mathbf{g}) = \ &\eta_i(\mathbf{g}_0) + \sum_j \left[\frac{\mathrm{d}\eta_i(\mathbf{g}_0)}{\mathrm{d}g_j} \right] \times g_j + \\
&\sum_j \frac{1}{2} \left[\frac{\mathrm{d}^2\eta_i(\mathbf{g}_0)}{\mathrm{d}(g_j)^2} \right] \times (g_j)^2 + \\
&\sum_{j,k} \frac{1}{2} \left[\frac{\mathrm{d}^2\eta_i(\mathbf{g}_0)}{\mathrm{d}g_j\mathrm{d}g_k} \right] \times g_j g_k + \\
&\sum_j \frac{1}{6} \left[\frac{\mathrm{d}^3\eta_i(\mathbf{g}_0)}{\mathrm{d}(g_j)^3} \right] \times (g_j)^3 + \cdots
\end{aligned}
\tag{8.8}
$$

Under the assumptions of a relatively small value of $\mathbf{g}$ and of well-behaved derivatives, the infinite series can be reduced by neglecting all second and higher order terms. This the yields the following approximation:

$$
\eta_i(\mathbf{g}_0 + \mathbf{g}) \approx \eta_i(\mathbf{g}_0) + \sum_j \left[\frac{\mathrm{d}\eta_i(\mathbf{g}_0)}{\mathrm{d}g_j} \right] \times g_j
\tag{8.9}
$$

This means that we may approximate the change in impact due to a relatively small additional intervention $\mathbf{g}$ as

$$
\Delta\eta_i(\mathbf{g}) \approx \sum_j \left[\frac{\mathrm{d}\eta_i(\mathbf{g}_0)}{\mathrm{d}g_j} \right] \times g_j
\tag{8.10}
$$

We then identify the change in impact $\Delta\eta_i$ as the impact that is attributed to an intervention vector $\mathbf{g}$. The above expression is a linear one, that can be rewritten as

$$
h_i(\mathbf{g}) \approx \mathbf{q}_i\mathbf{g}
\tag{8.11}
$$

where $\mathbf{q}_i$ represents the characterisation vector for impact category i. It is given as

$$
(\mathbf{q}_i)_j = \left[\frac{\mathrm{d}\eta_i(\mathbf{g}_0)}{\mathrm{d}g_j} \right]
\tag{8.12}
$$

Notice that $(\mathbf{q}_i)_j$ represents the characterisation factor for the contribution of intervention j to impact category i.

In the following sections, we will go on to consider the inventory vector

$$\mathbf{g} = \begin{pmatrix} 120 \\ 14 \\ -100 \end{pmatrix} \qquad (8.13)$$

We will assume that characterisation models η_1, η_2 and η_3 are available for modeling the effects on the impact categories acidification, global warming and resource scarcity, in terms of the basis

$$\begin{pmatrix} \text{kg SO}_2\text{-equivalent of H}^+\text{-release} \\ \text{kg CO}_2\text{-equivalent of infrared absorption} \\ \text{RDU of reduced availability} \end{pmatrix} \qquad (8.14)$$

Finally, we will assume that three vectors of characterisation factors have been derived from these three models, yielding:

$$\mathbf{q}_1 = \begin{pmatrix} 0 \\ 1 \\ 0 \end{pmatrix}, \mathbf{q}_2 = \begin{pmatrix} 1 \\ 0.1 \\ 0 \end{pmatrix} \text{ and } \mathbf{q}_3 = \begin{pmatrix} 0 \\ 0 \\ -15 \end{pmatrix} \qquad (8.15)$$

meaning, for instance, that of all environmental flows involved in the linear space, only sulphur dioxide contributes to acidification, and has a characterisation factor of 1 kg SO_2-equivalent of H^+-release per kg of sulphur dioxide. The negative coefficient in $\mathbf{q}_3$ implies that the extraction of crude oil, which is a negative intervention, leads to a positive contribution to resource scarcity.

8.1.3 Classification

Classification refers to the "assignment LCI results to impact categories" (ISO, 2000, p.7). This definition (and many other definitions from literature) leaves unclear whether only the inventory items are assigned as flow names, or that the quantities are assigned as well. For instance, given the fact that carbon dioxide is a greenhouse gas, and that the hypothetical product system of Section 2.2 shows an emission of 120 kg of carbon dioxide, one could assign either carbon dioxide to global warming, or 120 kg of carbon dioxide to global warming. Thus we may distinguish two variants of classification:

- qualitative assignment;

- quantitative assignment.

In both cases, the classification is based upon the question if a particular environmental intervention, say j, contributes to a particular impact category, say i. When all information in this respect is embodied in the characterisation factor $(\mathbf{q}_i)_j$, it means simply checking if this characterisation factor is zero or not. If it is zero, the interventions does not contribute to the impact category, if it is non-zero, it does. The Heaviside step function $\Theta(x)$ which is defined as

$$\Theta(x) = \begin{cases} 1 & \text{if } x > 0 \\ 0 & \text{otherwise} \end{cases} \tag{8.16}$$

provides a mathematical form of testing whether a variable is positive or not. Because non-zero characterisation factors may be positive or negative, we must use the Heaviside stepfunction of the absolute value:

$$\Theta(|x|) = \begin{cases} 1 & \text{if } x \neq 0 \\ 0 & \text{if } x = 0 \end{cases} \tag{8.17}$$

Classification according to the qualitative interpretation amounts to constructing

$$\hat{h}_{ij} = \Theta\left(\left|(\mathbf{q}_i)_j\right|\right) \Theta\left(|g_j|\right) \tag{8.18}$$

where $\hat{h}_{ij}$ is one of the results of the classification. It is 1 for a combination of environmental flow j and impact category i with a non-zero characterisation factor and a non-zero intervention, and it is 0 if either the characterisation factor or the intervention is zero (or both).

Classification according to the quantitative interpretation proceeds by

$$\hat{h}_{ij} = \Theta\left(\left|(\mathbf{q}_i)_j\right|\right) g_j \tag{8.19}$$

where $\hat{h}_{ij}$ is again one of the results of the classification. It is equal to the size of the intervention g_j when both characterisation factor and intervention differ from zero, and it is 0 if either the characterisation factor or the intervention is zero (or both).

An important function of the classification is to add information on poorly described environmental mechanisms. If certain environmental flows are known to contribute to a certain impact category but a characterisation factor is lacking, the classification can be used to include such information. In that case, the criterion used in classification is of course not a mechanical selection according to $\Theta\left(\left|(\mathbf{q}_i)_j\right|\right)$, but involves accounting of missing values in a way that characterisation factors may be non-zero, zero, or unknown.

Slightly related to this is that characterisation factors, derived according to Section 8.1.2, may be zero even when an environmental flow contributes to an impact category. A zero characterisation factor may arise when a characterisation model $\eta_i(\mathbf{g})$ has a maximum or minimum for g_j. For instance, when phosphate is a limiting factor for eutrophication, a small change of the emission of sulphur will not affect the level of eutrophication. Or, when all vegetation has disappeared due to excessively high levels of acid rain, the characterisation factor for nitrogen will be zero.

Sometimes, the problem of interventions with parallel or serial mechanisms is discussed in connection to classification. For instance, ISO (2000, p.7), states that "SO_2 is allocated between the impact categories of human health and acidification." In the qualitative interpretation of classification, this allocation problem is absent, as flows are assigned without caring about quantitative details. In the quantitative interpretation of classification, one would need factors for allocating a fraction of each flow to the different impact categories, *e.g.*,

$$\hat{h}_{ij} = \Xi_{ij} \Theta \left(\left\| (\mathbf{q}_i)_j \right\| \right) g_j \tag{8.20}$$

There is no guidance in how to construct such allocation factors Ξ_{ij}.

The result of the classification is a classification matrix $\hat{\mathbf{H}}$ with rows representing environmental flows and columns representing impact categories. Observe that this matrix differs from the "results of classification may be presented in a matrix" as described by Consoli *et al.* (1993, p.23). Also observe that, while characterisation often leads to a reduction of the number of variables, classification leads to an increase. No aggregation of is involved, although Fava *et al.* (1993, p.16) suggest that classification includes the "initial aggregation of data."

For the example, we find

$$\hat{\mathbf{H}} = \begin{pmatrix} 0 & 1 & 0 \\ 1 & 1 & 0 \\ 0 & 0 & 1 \end{pmatrix} \tag{8.21}$$

for qualitative classification, and

$$\hat{\mathbf{H}} = \begin{pmatrix} 0 & 14 & 0 \\ 120 & 14 & 0 \\ 0 & 0 & -100 \end{pmatrix} \tag{8.22}$$

for quantitative classification, or

$$\hat{\mathbf{H}} = \begin{pmatrix} 0 & 7 & 0 \\ 120 & 7 & 0 \\ 0 & 0 & -100 \end{pmatrix} \tag{8.23}$$

for quantitative classification with an allocation of sulphur dioxide over acidification and global warming on the basis of 0.5 and 0.5.

8.1.4 Characterisation

Characterisation "involves the conversion of LCI results to common units and the aggregation of the converted results within the impact categories" (ISO, 2000, p.7). Conversion is implied to mean multiplication. A general formula for characterisation is

$$h_i = \sum_j (\mathbf{q}_i)_j \, g_j \tag{8.24}$$

or

$$h_i = \mathbf{q}_i \mathbf{g} \tag{8.25}$$

When the characterisation vectors for several impact categories are juxtaposed, we can form the characterisation matrix $\mathbf{Q}$:

$$\mathbf{Q} = \begin{pmatrix} \mathbf{q}_1 & | & \mathbf{q}_2 & | & \cdots \end{pmatrix} \tag{8.26}$$

The formula for the characterisation into the various impact categories assumes the form

$$\mathbf{h} = \mathbf{Q}\mathbf{g} \tag{8.27}$$

We will refer to $\mathbf{h}$ as the impact vector.

The example leads to a characterisation matrix

$$\mathbf{Q} = \begin{pmatrix} 0 & 1 & 0 \\ 1 & 0.1 & 0 \\ 0 & 0 & -15 \end{pmatrix} \tag{8.28}$$

As a side remark, one may notice that this matrix is square. In general, this will no be so; in practice the matrix will have hundreds of rows and only ten or twenty columns. Because no inversion is needed, a non-square matrix is not of concern here. With

$$\mathbf{g} = \begin{pmatrix} 120 \\ 14 \\ -100 \end{pmatrix} \tag{8.29}$$

one finds

$$\mathbf{h} = \begin{pmatrix} 14 \\ 134 \\ 1500 \end{pmatrix} \tag{8.30}$$

for the impact vector.

From a computational point of view, this is all there is to say on characterisation. Some remarks apply, however. The inventory vector contain negative elements for inputs. Natural resources are an important example here. In the definition of the characterisation factors as the change in impact as a result of a small change in intervention in the positive direction, we will find that the characterisation factor that measures the contribution of crude oil to resource depletion is a negative number: a positive change means less extraction and therefore less depletion. The product of a negative element of $\mathbf{g}$ and a negative element of $\mathbf{Q}$ is positive, hence the converted intervention is positive, as it is for 'normal' pollutants. Occasionally, there may be negative characterisation factors for emitted substances as well. This is because some substances have a moderating effect on a certain type of impact. Finally, emissions may sometimes be negative themselves, as in the case of uptake of carbon dioxide by agriculture, and when the substitution method is used to allocate multifunctional processes so that the system contains 'avoided processes' (see Section 3.2.2). Such negative interventions lead, when multiplied with a positive characterisation factor, to a converted intervention with a negative sign.

8.1.5 Normalisation

Normalisation refers to the act of dividing the results of the characterisation by a suitable reference value. An often used reference value is the magnitude of the characterisation results for a reference area during a reference period, *e.g.*, the world during one year. We will indicate the reference inventory vector by $\dot{\mathbf{g}}$ and write

$$\dot{\mathbf{h}} = \mathbf{Q}\dot{\mathbf{g}} \tag{8.31}$$

to indicate the reference impact vector $\dot{\mathbf{h}}$. Notice the use of the dot, as a mnemonic for the fact that product-related interventions are usually measured in terms of kg, while the interventions for a reference period are measured in terms of kg/yr.

The normalisation step itself now consists of dividing each element of $\mathbf{h}$ by the corresponding element of $\dot{\mathbf{h}}$ to yield a normalised indicator result for impact category i:

$$\tilde{h}_i = \frac{h_i}{\dot{h}_i} \tag{8.32}$$

which obviously makes sense only when the reference value is non-zero. This is an unproblematic restriction, because one of the reasons to consider a specific impact category is the fact that it is a non-trivial issue. Normally, elements of the vector $\mathbf{h}$ have units such as kg CO_2-equivalent, and elements of $\dot{\mathbf{h}}$ have units such as kg CO_2-equivalent/yr. The elements of the normalised impact vector $\tilde{\mathbf{h}}$ is the ratio between these two, and is normally expressed in time units, such as yr.

If, in the hypothetical example, the annual environmental flows are given by

$$\dot{\mathbf{g}} = \begin{pmatrix} 10^{11} \\ 5 \times 10^{10} \\ -10^9 \end{pmatrix} \tag{8.33}$$

we find

$$\dot{\mathbf{h}} = \begin{pmatrix} 5 \times 10^{10} \\ 1.1 \times 10^{11} \\ 1.5 \times 10^{10} \end{pmatrix} \tag{8.34}$$

so that

$$\tilde{\mathbf{h}} = \begin{pmatrix} 2.8 \times 10^{-10} \\ 1.2 \times 10^{-9} \\ 1 \times 10^{-7} \end{pmatrix} \tag{8.35}$$

is the normalised impact vector.

8.1.6   Grouping*

ISO (2000, p.9) defines grouping as "assigning impact categories into one or more sets". Although have not seen many proposals for or examples of grouping, it seems to involve the partitioning of the impact categories, or probably the numerical results for all every impact categories $\mathbf{h}$, into a small number of sets. This is discussed towards the end of the section on contribution analysis (8.2.1).

8.1.7   Weighting

Weighting is a controversial issue. It requires the condensation of subjective opinion into quantitative measures that may be used for decision-making.

From a computational point of view, weighting is simple. Given a weighting vector $\mathbf{w}$, where element w_i provides the weighting factor for impact category i, weighting proceeds by

$$W = \sum_i w_i \tilde{h}_i \qquad (8.36)$$

or in matrix form

$$W = \mathbf{w}\tilde{\mathbf{h}} \qquad (8.37)$$

Here W represents the weighted index. Usually, weights are dimensionless, so the weighted index agrees in dimension with the elements of $\tilde{\mathbf{h}}$; normally it will be yr. Observe that we have restricted ourselves to a linear weighting model; see, however, Section 9.1 for some extensions.

Occasionally, weighting may proceed into more than one index, for instance one for the area of protection human health and one for the area of protection ecosystem health. In those cases, one would write

$$W_k = \sum_i w_{ki} \tilde{h}_i \qquad (8.38)$$

where w_{ki} is the weighted index for area of protection k.

It may also happen that the weights do not relate to normalised scores but to the characterisation results directly. In that case, one obviously has

$$W = \sum_i w_i h_i \qquad (8.39)$$

Notice that strict requirements apply in this case due to the fact that the elements of $\mathbf{h}$ have in general differing units or even dimensions. The units should be equalised by means of the correspondingly different units of the elements of $\mathbf{w}$.

For the example, we work with a hypothetical weighting vector

$$\mathbf{w} = \begin{pmatrix} 8 \\ 15 \\ 3 \end{pmatrix} \qquad (8.40)$$

which gives

$$W = 3.2 \times 10^{-7} \qquad (8.41)$$

for the weighted index.

8.2   Interpretation

In most texts on LCA, it is acknowledged that goal and scope definition, inventory analysis and impact assessment deal with uncertain information. Nevertheless, the results of these phases are most often expressed as point estimates. The fourth phase, interpretation, is then used to address uncertainties and their influence in depth. Uncertainty analyses, the study of alternative scenarios, the influence of choices and the significance of data gaps are approached in this phase. In this book, we have chosen to discuss perturbation theory and the influence of uncertainties in a prior chapter (6). This means that a large part of the material has already been discussed. In this section, we still reserve a place for a short discussion of uncertainty analysis, to bring together all relevant information on that topic at one systematic place. Certain other aspects of interpretation are discussed as well. We largely follow Heijungs & Kleijn (2001) as to the choice of topics addressed. This means that all types of analysis of a procedural character are considered to be outside the topic of this book.

8.2.1   Contribution analysis

When the discussion is confined to computational aspects, inventory results are summarised in the inventory vector $\mathbf{g}$. The formula for it was derived in Section 2.2:

$$\mathbf{g} = \mathbf{B}\mathbf{A}^{-1}\mathbf{f} \tag{8.42}$$

which is a concise notation for

$$\forall k : g_k = \sum_{\forall j} \sum_{\forall i} b_{kj} \left(\mathbf{A}^{-1}\right)_{ji} f_i \tag{8.43}$$

Here the sum over j represents the aggregation over all unit processes, and the sum over i the aggregation over all economic flows that link these processes. Notice that we cannot employ the intensity matrix in the above equation, as its elements would only have the subscripts for ki; the summation over j is implicitly carried out in the intensity matrix. This means that LCA on the basis of pre-aggregated inventories (see Section 3.8.2) is incompatible with the sort of contribution analysis that is described below.

There are situations in which only a partial aggregation or no aggregation at all are desired. Such situations include:

- to investigate the contribution of a unit process, *e.g.*, production of ethylene, to a particular intervention, *e.g.*, the emission of CO_2;

- to investigate the contribution of a set of unit processes, *e.g.*, those belonging to the materials production phase, to a particular intervention, *e.g.*, the emission of CO_2.

It is appropriate to introduce the term partial intervention to indicate a particular environmental flow for which the double summation has not been executed fully. Let us partition the set of unit processes P into a number of smaller sets $P_1, P_2, \ldots$ such that

$$\bigcup_a P_a = P \tag{8.44}$$

and

$$\forall a \neq b : P_a \cap P_b = \emptyset \tag{8.45}$$

Now, denote the partial intervention for the set of unit processes P_a by

$$\forall k : g_k(P_a) = \sum_{j \in P_a} \sum_{\forall i} b_{kj} \left(A^{-1}\right)_{ji} f_i \tag{8.46}$$

Because the subsets form a partition of the full set, we have

$$\forall k : \sum_a g_k(P_a) = g_k \tag{8.47}$$

so that we may indeed interpret $g_k(P_a)$ as the partial intervention for environmental flow k. We can interpret this further as a contribution to the total intervention g_k when we form the ratio $\dfrac{g_k(P_a)}{g_k}$ so that

$$\forall k : \sum_a \frac{g_k(P_a)}{g_k} = 1 \tag{8.48}$$

In connection with the substitution method for allocating multifunctional processes (see Section 3.2.2), it was remarked that emission may be negative, because they are emissions from an avoided process. When a certain substance is emitted by certain normally included processes as well as by certain avoided processes, certain problems in interpreting a contribution analysis may occur. It may, for instance, happen that a total emission is positive, say 20 kg, and that one partition of processes yields from a contribution analysis 30 kg while the other partition of processes yields -10 kg. It seems strange to state that the first partition contributes 150% while the second partition contributes -50%. It may even be the case that normally included processes and avoided processes exactly cancel one another for a certain emission, for instance contributing 10 kg and -10 kg. In that case, one cannot even specify contribution as percentages.

It is not always simple to define a sensible partition of the set of unit processes. Moreover, a unit process is fully assigned to one and only one subset of unit processes. This means, for instance, that if the unit process of production of electricity delivers electricity to the processes of production of polyethylene and using a refrigerator, it is not divided among a subset materials production and a subset household activities. It can be included in one of them, or it can be included in an entirely different subset (like energy production).

The most detailed partition that can be formed is one in which each subset P_a contains only one unit process, *e.g.*, process a. The advantage of this fine partitioning that the aforementioned problem of forming sensible partitions is omitted. The contribution analysis than simply studies the contributions to a certain intervention by each individual process.

We may observe that the inventory vector follows from a double summation, and that the partial intervention has been discussed only in relation to one of these two summations. The second summation is over economic flows (index i). In principle, it is possible to define a partition of the set of economic flows E into a number of smaller sets $E_1, E_2, \ldots$ with identical rules for their union and cross-section. Another partial intervention for the set of economic flows E_a can be formed by

$$\forall k : g_k(E_a) = \sum_{\forall j} \sum_{i \in E_a} b_{kj} \left(\mathbf{A}^{-1}\right)_{ji} f_i \tag{8.49}$$

Recall, however, that in most cases the final demand vector $\mathbf{f}$ is a vector with zeros everywhere except at one place, the flow that corresponds to the reference flow of the system. This type of decomposition is therefore of little practical interest in LCA. In a more scenario-based analysis of economy-wide consumption patterns, a composite final demand vector may be interesting; see Section 3.9.3. In that case, contribution by several subsets of economic flows may be a topic of study, for instance in decomposing a household's impacts into the food-related part and the non food-related part.

Going beyond the inventory analysis, we proceed to the characterisation. The formula for the impact vector can be written as

$$\mathbf{h} = \mathbf{QBA}^{-1}\mathbf{f} \tag{8.50}$$

which stands for

$$\forall l : h_l = \sum_{\forall k} \sum_{\forall j} \sum_{\forall i} q_{lk} b_{kj} \left(\mathbf{A}^{-1}\right)_{ji} f_i \qquad (8.51)$$

Here the sum over k represents the aggregation over all environmental interventions, and q_{lk} the characterisation factor linking intervention k and impact category l. There are now three summations, and when we ignore decomposition of the sum over economic flows i, there remain two directions of decomposition:

- a contribution analysis according to unit processes;

- a contribution analysis according to environmental interventions.

For the decomposition in terms of processes, we extend from before and obtain a partial impact score

$$\forall l : h_l(\mathsf{P}_a) = \sum_{\forall k} \sum_{j \in \mathsf{P}_a} \sum_{\forall i} q_{lk} b_{kj} \left(\mathbf{A}^{-1}\right)_{ji} f_i \qquad (8.52)$$

For the decomposition in terms of interventions, we may repeat the procedure of defining a partition for interventions. We may partition the set of interventions I into a number of smaller sets $\mathsf{I}_1, \mathsf{I}_2, \ldots$ with the by now familiar rules for union and cross-section. This then may be used to define the partial impact score

$$\forall l : h_l(\mathsf{I}_a) = \sum_{k \in \mathsf{I}_a} \sum_{\forall j} \sum_{\forall i} q_{lk} b_{kj} \left(\mathbf{A}^{-1}\right)_{ji} f_i \qquad (8.53)$$

The two directions for decomposition may also be followed at the same time, yielding

$$\forall l : h_l(\mathsf{P}_a, \mathsf{I}_b) = \sum_{k \in \mathsf{I}_b} \sum_{j \in \mathsf{P}_a} \sum_{\forall i} q_{lk} b_{kj} \left(\mathbf{A}^{-1}\right)_{ji} f_i \qquad (8.54)$$

For this latter form, we have

$$\forall l : \sum_a \sum_b h_l(\mathsf{P}_a, \mathsf{I}_b) = h_l \qquad (8.55)$$

When the number of subsets along which P is partitioned and the number of subsets along which I is partitioned are both large, the decomposition takes place along very many subsets. This may in some cases be problematic, for reasons of interpreting a very long list, and because most individual contributions will become very small.

In most practical applications, decomposition of interventions is carried out at the level of individual interventions, not at that of subsets that contain more than one element. However, such larger subsets of interventions may provide sensible partitions, *e.g.*, when one investigates the contribution to toxicity by all pesticides or by all persistent chemicals.

The remark on problems in interpreting negative contributions is also applicable at this place. Occasionally, characterisation factors are negative, *e.g.*, certain chemicals prohibit photo-oxidant formation. When analysing the contributions by different interventions to photo-oxidant formation, most chemicals will therefore show a zero or positive contribution while a number of them will show a negative contribution.

The level of normalisation does not add more directions for decomposition. The formula is

$$\forall l : \tilde{h}_l = \frac{1}{\tilde{h}_l} \sum_{\forall k} \sum_{\forall j} \sum_{\forall i} q_{lk} b_{kj} \left(\mathbf{A}^{-1} \right)_{ji} f_i \qquad (8.56)$$

so that the discussion on characterisation can be repeated here.

There is one additional direction at the level of weighting. The formula is

$$W = \sum_{\forall l} \sum_{\forall k} \sum_{\forall j} \sum_{\forall i} \frac{w_l}{\tilde{h}_l} q_{lk} b_{kj} \left(\mathbf{A}^{-1} \right)_{ji} f_i \qquad (8.57)$$

Defining a partition of the set of impact categories C into a number of smaller sets $\mathsf{C}_1, \mathsf{C}_2, \dots$ provides a third direction for decomposition. We write

$$W(\mathsf{C}_a) = \sum_{l \in \mathsf{C}_a} \sum_{\forall k} \sum_{\forall j} \sum_{\forall i} \frac{w_l}{\tilde{h}_l} q_{lk} b_{kj} \left(\mathbf{A}^{-1} \right)_{ji} f_i \qquad (8.58)$$

to indicate the contribution of the set of impact categories C_a to the weighted index. The other two directions are given by

$$W(\mathsf{I}_a) = \sum_{\forall l} \sum_{k \in \mathsf{I}_a} \sum_{\forall j} \sum_{\forall i} \frac{w_l}{\tilde{h}_l} q_{lk} b_{kj} \left(\mathbf{A}^{-1} \right)_{ji} f_i \qquad (8.59)$$

and

$$W(\mathsf{P}_a) = \sum_{\forall l} \sum_{\forall k} \sum_{j \in \mathsf{P}_a} \sum_{\forall i} \frac{w_l}{\tilde{h}_l} q_{lk} b_{kj} \left(\mathbf{A}^{-1} \right)_{ji} f_i \qquad (8.60)$$

The three directions may be combined in four different ways:

$$W(\mathsf{P}_a, \mathsf{C}_b) = \sum_{l \in \mathsf{C}_b} \sum_{\forall k} \sum_{j \in \mathsf{P}_a} \sum_{\forall i} \frac{w_l}{h_l} q_{lk} b_{kj} \left(\mathbf{A}^{-1}\right)_{ji} f_i \qquad (8.61)$$

for the contributions by processes and impact categories,

$$W(\mathsf{I}_a, \mathsf{C}_b) = \sum_{l \in \mathsf{I}_a} \sum_{k \in \mathsf{C}_b} \sum_{\forall j} \sum_{\forall i} \frac{w_l}{h_l} q_{lk} b_{kj} \left(\mathbf{A}^{-1}\right)_{ji} f_i \qquad (8.62)$$

for the contributions by interventions and impact categories,

$$W(\mathsf{P}_a, \mathsf{I}_b) = \sum_{\forall l} \sum_{k \in \mathsf{I}_b} \sum_{j \in \mathsf{P}_a} \sum_{\forall i} \frac{w_l}{h_l} q_{lk} b_{kj} \left(\mathbf{A}^{-1}\right)_{ji} f_i \qquad (8.63)$$

for the contributions by processes and interventions, and

$$W(\mathsf{P}_a, \mathsf{I}_b, \mathsf{C}_c) = \sum_{l \in \mathsf{C}_c} \sum_{k \in \mathsf{I}_b} \sum_{j \in \mathsf{P}_a} \sum_{\forall i} \frac{w_l}{h_l} q_{lk} b_{kj} \left(\mathbf{A}^{-1}\right)_{ji} f_i \qquad (8.64)$$

for the contributions by processes, interventions and impact categories. Notice that we have left out the decomposition in the fourth direction – that of economic flows – in the entire discussion. when meaningful, it may be added in a trivial way.

Partitions of the set of impact categories are not often made, apart from treating each impact category as one such subset. Possible sensible subsets could comprise: all toxicity-related impact categories, all chemical-related impact categories, and so on. This variant of contribution analysis is probably the same as what ISO (2000) has in mind with grouping (see Section 8.1.6).

Contribution analyses have been performed in many case studies. Their theoretical foundation has not received much attention, however. An exception is provided by Noh *et al.* (1998).

8.2.2 Structural analysis

Chapter 7 discussed the tentative principles of structural theory in connection to the inventory analysis. Now, we are ready to assign its application a place in the interpretation phase, under the name of structural analysis. Structural analysis could then be defined as the systematic study of the presence of patterns in LCA data. As the most important object for structural analysis, the intensity matrix has been proposed for analysis in correlation analysis (Section 7.2.3) as well as in principal component analysis (Section 7.2.4). Because the details and usefulness of these techniques are not yet clear, we will restrict ourselves here to this brief reference. In addition, we may mention the quantities that can be used analogous to the intensity matrix at various levels:

- the inverse of the technology matrix $\mathbf{A}^{-1}$ at the level of the scaling vector;

- the intensity matrix $\mathbf{\Lambda}$ at the level of the inventory vector;

- the matrix $\mathbf{Q\Lambda}$ at the level of the characterisation.

There is no need for considering the level of normalisation here. This is because PCA starts from the covariance matrix, so that changes of scale introduced by normalisation tend to be ignored. The level of weighting has been left out, because weighting yields data in a one-dimensional space, that cannot be further reduced by PCA or similar techniques.

8.2.3   Perturbation analysis

In Section 6.1, we discussed the sensitivity of the solution of a system for small perturbations in the coefficients of the equations. We were lead to consider quantities like $\dfrac{\partial g_k}{\partial a_{ij}}$ and $\dfrac{\partial g_k}{\partial b_{ij}}$, and derived algebraic expressions for these. In a perturbation analysis, the emphasis is not so much on the absolute but on the relative sensitivity. That is, a typical question could be: 'If a certain process coefficient changes by 1%, how many % change the environmental interventions?' A key to the answers is to be found in the first-order approximations for δg_k:

$$\delta g_k \approx \frac{\partial g_k}{\partial a_{ij}} \delta a_{ij} \text{ and } \delta g_k \approx \frac{\partial g_k}{\partial b_{ij}} \delta b_{ij} \tag{8.65}$$

This leads to

$$\frac{\delta g_k}{g_k} \approx \frac{a_{ij}}{g_k} \frac{\partial g_k}{\partial a_{ij}} \frac{\delta a_{ij}}{a_{ij}} \text{ and } \frac{\delta g_k}{g_k} \approx \frac{b_{ij}}{g_k} \frac{\partial g_k}{\partial b_{ij}} \frac{\delta b_{ij}}{b_{ij}} \tag{8.66}$$

Such a question thus calls for a study of a number of quantities γ_k defined by

$$\gamma_k(a_{ij}) = \frac{\partial g_k/g_k}{\partial a_{ij}/a_{ij}} \text{ and } \gamma_k(b_{ij}) = \frac{\partial g_k/g_k}{\partial b_{ij}/b_{ij}} \tag{8.67}$$

which approximate the question mentioned as can be seen from

$$\frac{\delta g_k}{g_k} \approx \gamma_k(a_{ij})\frac{\delta a_{ij}}{a_{ij}} \text{ and } \frac{\delta g_k}{g_k} \approx \gamma_k(b_{ij})\frac{\delta b_{ij}}{b_{ij}} \tag{8.68}$$

We will indicate the matrix formed for one k and several values of i and j by

$$\boldsymbol{\Gamma}_k(\mathbf{A}) = \begin{pmatrix} \dfrac{\partial g_k/g_k}{\partial a_{11}/a_{11}} & \dfrac{\partial g_k/g_k}{\partial a_{12}/a_{12}} & \cdots \\ \dfrac{\partial g_k/g_k}{\partial a_{21}/a_{21}} & \dfrac{\partial g_k/g_k}{\partial a_{22}/a_{22}} & \cdots \\ \cdots & \cdots & \cdots \end{pmatrix} \tag{8.69}$$

and

$$\boldsymbol{\Gamma}_k(\mathbf{B}) = \begin{pmatrix} \dfrac{\partial g_k/g_k}{\partial b_{11}/b_{11}} & \dfrac{\partial g_k/g_k}{\partial b_{12}/b_{12}} & \cdots \\ \dfrac{\partial g_k/g_k}{\partial b_{21}/b_{21}} & \dfrac{\partial g_k/g_k}{\partial b_{22}/b_{22}} & \cdots \\ \cdots & \cdots & \cdots \end{pmatrix} \tag{8.70}$$

and refer to these matrices as perturbation matrices for the inventory vector. We need not worry too much about what happens when g_k, a_{ij} or b_{ij} is zero: $g_k = 0$ denotes an intervention that is not involved, and we may exclude it from the analysis, and $a_{ij} = 0$ and $b_{ij} = 0$ can lead us to define $\gamma_k = 0$.

Inserting the expressions for $\dfrac{\partial g_k}{\partial a_{ij}}$, $\dfrac{\partial g_k}{\partial b_{ij}}$ and $\dfrac{\partial g_k}{\partial f_i}$ derived in Section 6.1, we find

$$\gamma_k(a_{ij}) = -\frac{a_{ij}}{g_k}\lambda_{ki}s_j \tag{8.71}$$

and

$$\gamma_k(b_{ij}) = \begin{cases} -\dfrac{b_{ij}}{g_k}\lambda_{ki}s_j & \text{if } i = k \\ 0 & \text{otherwise} \end{cases} \tag{8.72}$$

We may also derive an expression for $\gamma_k(f_i) = \dfrac{\partial g_k/g_k}{\partial f_i/f_i}$:

$$\gamma_k(f_i) = \frac{f_i}{g_k}\lambda_{ki} \tag{8.73}$$

Because $g_k = \displaystyle\sum_j \lambda_{ki}f_i$ and f_i is most often 0 for all but one i, we usually have

$$\gamma_k(f_i) = \begin{cases} 1 & \text{if } i = r \\ 0 & \text{otherwise} \end{cases} \tag{8.74}$$

In the example system, we have

$$\Gamma_1(\mathbf{A}) = \begin{pmatrix} 0.17 & -0.17 \\ -1 & 0 \end{pmatrix} \tag{8.75}$$

and

$$\Gamma_2(\mathbf{A}) = \begin{pmatrix} 0.29 & -0.29 \\ -1 & 0 \end{pmatrix} \tag{8.76}$$

and

$$\Gamma_3(\mathbf{A}) = \begin{pmatrix} 1 & -1 \\ -1 & 0 \end{pmatrix} \tag{8.77}$$

for perturbations of the technology matrix $\mathbf{A}$, and

$$\Gamma_1(\mathbf{B}) = \begin{pmatrix} 0.83 & 0.17 \\ 0 & 0 \\ 0 & 0 \end{pmatrix} \tag{8.78}$$

and

$$\Gamma_2(\mathbf{B}) = \begin{pmatrix} 0 & 0 \\ 0.71 & 0.29 \\ 0 & 0 \end{pmatrix} \tag{8.79}$$

and

$$\Gamma_3(\mathbf{B}) = \begin{pmatrix} 0 & 0 \\ 0 & 0 \\ 0 & 1 \end{pmatrix} \tag{8.80}$$

for perturbations of the intervention matrix $\mathbf{B}$.

For a given environmental flow k, one may compute all possible values of γ_k, and tabulate them in ascending or descending order. Here, one may decide to mix up $\gamma_k(a_{ij})$ and $\gamma_k(b_{ij})$ in one table. One may wish to ignore the sign in the sorting procedure. One may also wish to ignore values of γ_k that are small enough to be uninteresting, e.g. those for which $|\gamma_k| < 0.5$. It should be noted that Sebald (1974) discusses a similar concept for input-output analysis, under the name "most important parameter problem." Heijungs (1994) introduces it in LCA under the name "marginal analysis." The quantities γ_k are known as multipliers, tolerance amplifications, or amplification factors. For the example, we find the table below.

We may note that for certain combinations of i, j and k, $\gamma_k(a_{ij})$ is positive and for others negative. This is interpreted in terms of the direction of sensitivity. If $\gamma_k(a_{ij})$ is negative, a small increase of a_{ij} leads to a decrease of g_k.

Table 8.1: Sorted multipliers for g_1 (carbon dioxide) for a perturbation analysis of the example case.

Coefficient	Multiplier
a_{21}	1
b_{11}	0.83
a_{11}	0.17
a_{12}	-0.17
b_{11}	0.17

We may also note that it may occur that certain values of $\gamma_k(a_{ij})$ are larger than 1 or smaller than -1. This points out that these parameters are quite sensitive to perturbations. A value of 2 means that a change of a_{ij} by 1% induces a change in g_k by approximately 2%. It is especially these values that are of interest in iterative improvement of data reliability and product and process improvement. It is therefore an interesting question when $|\gamma| > 1$. There is, unfortunately, no easy answer to this. We may approach it by considering the relative sensitivity of the scaling factors for perturbations of the technology matrix, contained in the matrix $\Sigma_k(\mathbf{A})$:

$$\sigma_k(a_{ij}) = \frac{\partial s_k/s_k}{\partial a_{ij}/a_{ij}} = \frac{a_{ij}}{s_k}\frac{\partial s_k}{\partial a_{ij}} = -\frac{a_{ij}}{s_k}\left(\mathbf{A}^{-1}\right)_{ki} s_j \tag{8.81}$$

Assuming the usual case of only one non-zero reference flow, $f_r = \phi$, we obtain

$$\sigma_k(a_{ij}) = -a_{ij}\frac{\left(\mathbf{A}^{-1}\right)_{jr}\left(\mathbf{A}^{-1}\right)_{ki}}{\left(\mathbf{A}^{-1}\right)_{kr}} \tag{8.82}$$

There does not appear a clue to the question under which conditions $|\sigma_k(a_{ij})| > 1$. The conditions under which $|\gamma_k(a_{ij})| > 1$ or $|\gamma_k(b_{ij})| > 1$ are even more obscure. Experience suggests that a prerequisite for $|\gamma_k(a_{ij})| > 1$ is the existence of feedback loops in the technology matrix.

We may finally notice that the sum rows of Γ_k appears to be 0 or ± 1. We can verify this from

$$\sum_j \gamma_k(a_{ij}) = -\frac{\lambda_{ki}}{g_k}\sum_j a_{ij}s_j = -\frac{\lambda_{ki}}{g_k}f_i = -\gamma_k(f_i) \tag{8.83}$$

In the most usual case, with $f_i = 0$ for all i except when $i = r$, this leads to

$$\sum_j \gamma_k(a_{ij}) = \begin{cases} -1 & \text{if } i = r \\ 0 & \text{otherwise} \end{cases} \tag{8.84}$$

Furthermore,

$$\sum_j \gamma_k(b_{ij}) = \begin{cases} \dfrac{1}{g_k} \sum_j b_{ij} s_j = 1 & \text{if } i = k \\ 0 & \text{otherwise} \end{cases} \tag{8.85}$$

Thus, the sum rows is 0 or -1 for the $\mathbf{\Gamma}_k(\mathbf{A})$ and 0 or 1 for the $\mathbf{\Gamma}_k(\mathbf{B})$.

Similar to $\gamma(a)$ and $\gamma(b)$ as a measure of the perturbation in g, other measures may be constructed for:

- scaling factors: $\sigma(a)$, where σ measures the change in s;

- characterisation: $\eta(a), \eta(b), \eta(q)$, where η measures the change in h;

- normalisation: $\tilde{\eta}(a), \tilde{\eta}(b), \tilde{\eta}(q), \tilde{\eta}(\dot{g})$, where $\tilde{\eta}$ measures the change in $\tilde{h}$;

- weighting: $\omega(a), \omega(b), \omega(q), \omega(\dot{g}), \omega(w)$, where ω measures the change in W.

Observe that the number of parameters that may perturbed increases with the level of aggregation. There are more parameters that may be perturbed; one may think of allocation factors and exchange factors.

Instead of using the analytical formulae for the perturbation analysis, a numerical approach may be used as well. In that case, one forms

$$\frac{\partial g_k}{\partial a_{ij}} \approx \frac{g_k(a_{ij} + \epsilon) - g_k(a_{ij})}{\epsilon} \tag{8.86}$$

and takes a non-vanishing but small value for ϵ, *e.g.*, 0.01.

8.2.4   Uncertainty analysis

As discussed in Section 6.4, Monte Carlo simulations are a powerful tool in studying the uncertainties of the results of an LCA. Let $(\mathbf{g}^i)_k$ denote environmental intervention k in Monte Carlo trial number i. A number of Monte Carlo trials, say N then yields a set of values $\{(\mathbf{g}^1)_k, (\mathbf{g}^2)_k, \cdots, (\mathbf{g}^N)_k\}$. This set may be subject to statistical analysis. Statistics of interest include:

- the mean value $\bar{g}_k = \dfrac{1}{N} \sum_{i=1}^{N} (\mathbf{g}^i)_k$;

- the variance $s^2(g_k) = \dfrac{1}{N-1} \sum\limits_{i=1}^{N} \left((\mathbf{g}^i)_k - \bar{g}_k \right)^2$, the standard devia-

  tion $s(g_k) = \sqrt{s^2(g_k)}$ or the coefficient of variation $CV(g_k) = \dfrac{s(g_k)}{\bar{g}_k}$;

- the highest value $g_k^+ = \max\limits_{i=1}^{N} \left(\mathbf{g}^i \right)_k$ and lowest value $g_k^- = \min\limits_{i=1}^{N} \left(\mathbf{g}^i \right)_k$.

Under the assumption of a Gaussian distribution, a confidence interval may be constructed as $[\bar{g}_k - t(\alpha, N-1)s(g_k), \bar{g}_k + t(\alpha, N-1)s(g_k)]$ with the critical value of the t-distribution at significance level α and degrees of freedom $N-1$. With the conventional $\alpha = 0.95$ and sufficiently large N, this factor converges to 1.96.

Other statistics are measures of skewness and kurtosis, and the Kolmogorov-Smirnov test for normality. See, *e.g.*, Hays (1988) for a discussion. In addition, a histogram may reveal clues to the frequency distribution.

Obviously, the environmental intervention g may be replaced by s, h, $\tilde{h}$ or W for uncertainty analyses at the level of scaling factors, characterisation, normalisation and weighting.

8.2.5   Key issue analysis

Following Heijungs (1996), uncertainty analysis and perturbation analysis may be combined into an analysis of key issues. This is not key issues in the sense of Noh *et al.* (1998), who use this term for "a unit process and inventory item of which potential impact to the environment is significant within a given product system," (p.59–60) and which has been discussed under contribution analysis above. Rather, it refers to "areas to concentrate on in a more detailed LCI" (Heijungs (1996, p.160)). As such, it is a search for parameters that are uncertain and for which the result is sensitive. See Table 8.2 for a categorisation of parameters into three sets: "a key issue," "not a key issue," and "perhaps a key issue."

The formula for error propagation that was derived in Section 6.2

$$\sigma^2(g_k) = \sum_{i,j} \lambda_{ki}^2 s_j \sigma^2(a_{ij}) \tag{8.87}$$

can be used for key issue identification. One then forms expression like

$$\zeta_{ij} = \frac{\sum\limits_{i,j} \lambda_{ki}^2 s_j \sigma^2(a_{ij})}{\sigma^2(g_k)} \tag{8.88}$$

Table 8.2: Data that is certain and hardly contributes to the analysis ("not a key issue") must be separated from data that is uncertain and makes quite some contribution ("a key issue"). Anything in between must be considered carefully ("perhaps a key issue").

	contribution	
uncertainty	low	high
low	not a key issue	perhaps a key issue
high	perhaps a key issue	a key issue

where ζ_{ij} is the contribution to the overall variance that is made by the variance of a_{ij}. One can now figure out all these coefficients ζ_{ij} and label the largest ones as key issues.

8.2.6 Comparative analysis

The comparative analysis presents a certain environmental of several product alternatives simultaneously with the aim of offering insight with respect to strengths and weaknesses of each alternative. Let us concentrate on one environmental flow, say k. The comparative analysis then addresses $g_{k1}, g_{k2}, \ldots$. We may distinguish several useful ways of tabulating the scores for the different alternatives:

- according to absolute value: $g_{k1}, g_{k2}, \ldots$;

- with one alternative, say i, acting as a reference system, so that all scores are expressed relative to that system: $\dfrac{g_{k1}}{g_{ki}}, \dfrac{g_{k2}}{g_{ki}}, \ldots$;

- with the alternative with the highest score acting as a reference system: $\dfrac{g_{k1}}{\max_i g_{ki}}, \dfrac{g_{k2}}{\max_i g_{ki}}, \ldots$;

- with the alternative with the lowest score acting as a reference system: $\dfrac{g_{k1}}{\min_i g_{ki}}, \dfrac{g_{k2}}{\min_i g_{ki}}, \ldots$

All these possibilities are often seen in combination with a bar diagram. Comparative analyses may also be made at other levels than inventory analysis. At the level of characterisation, g may be replaced by h; at that of normalisation by $\tilde{h}$, and at that of weighting by W. One may even choose to compare scaling factors by selecting s.

In putting one of the alternatives to 1, we must be aware that g_{ki}, $\max_i g_{ki}$ or $\min_i g_{ki}$ may occasionally be zero, so that it cannot serve as a denominator. Also, care should be taken that in rare cases g_{ki} may range from negative to positive for various product alternatives, so that one may consider to use $\max_i |g_{ki}|$ or instead $\min_i |g_{ki}|$.

8.2.7 Discernibility analysis

Section 8.2.4 discusses the analysis of uncertainties for one product alternative. One may easily imagine that such an uncertainty analysis gives a 95%-confidence interval that ranges from 10 to 15 kg of carbon dioxide for a certain product alternative, and from 11 to 16 kg of carbon dioxide for a second product alternative. The confidence intervals largely overlap, and one might conclude that the two products are indiscernible in the sense of showing no significant difference as to carbon dioxide. This ignores one important aspect: it is not fair to compare the two products in two separate Monte Carlo trials in which, say, the performance of the electric power plant was good for product 1 and bad for product 2. It is much more reasonable to compare the products pair-wise in each Monte Carlo trial. Following Huijbregts (1998), one then computes the ratio of the carbon dioxide emission for the two products. The resulting distribution of the "comparison indictor" can be analysed as to the percentage of cases that it is smaller or larger than 1. A 95%-criterion could then be applied to distinguish significant from non-significant differences.

The approach becomes quite complex when several product alternative are studied. For 3 products, we would need to compute the ratio of the emission for product 1 to product 2 and for product 2 to product 3 for every Monte Carlo trial. So, the number of frequency distributions increases with the number of products compared.

One could solve this by reducing the frequency distributions to the first and second moment, *i.e.* the mean and variance. One may note that Huijbregts' (1998) idea of studying the distribution of the comparison index is in that case reducible to applying a t-test for paired samples. The generalisation of the t-test for more than two samples is the analysis of variance (ANOVA), and the fact that we have paired or related samples would induce one to apply a two-way analysis of variance without interaction. The disadvantage, however, is that this leads to an omnibus test (based on the f-statistic), in which the null hypothesis is that all products have the same mean score. Rejection of the hypothesis does not convey much, except that at least one product alternative has a significantly different score. See, for instance, Sheskin (1997) for background material on statistical tests.

A solution can be found by noting that, although the full frequency distribution is constructed, only the fraction of times that it exceeds 1 matters. The problem of comparing more than two products can be solved when we switch to another type of comparison index (Heijungs & Kleijn, 2001). In Section 3.4.2 and 3.8.1, we discussed the fact that one and the same combination of technology matrix and intervention matrix may be applied to different final demand vectors. This is of particular importance in the present simultaneous analysis of product alternatives in the presence of uncertain information. In one Monte Carlo trial, one constructs the two stochastic matrices $\mathbf{A}^i$ and $\mathbf{B}^i$, where the superscript i indicates that the matrices are particular realisations of a stochastic procedure in trial i. They are combined to form the intensity matrix $\mathbf{\Lambda}^i$. This matrix is then applied to all investigated product alternatives, described by the several final demand vectors $\mathbf{f}_1, \mathbf{f}_2, \mathbf{f}_3, \ldots$. This yields several inventory vectors $\mathbf{g}_1^i, \mathbf{g}_2^i, \mathbf{g}_3^i, \ldots$. Let us concentrate on one of the rows of these flows, say $\left(\mathbf{g}_1^i\right)_k, \left(\mathbf{g}_2^i\right)_k, \left(\mathbf{g}_3^i\right)_k, \ldots$, e.g., representing carbon dioxide. What ultimately matters in a comparative analysis is not the size of the difference, but the direction only: is $\left(\mathbf{g}_1^i\right)_k$ larger than $\left(\mathbf{g}_2^i\right)_k$, is it smaller, or are they equal? So the analysis focuses on the difference $\left(\mathbf{g}_1^i\right)_k - \left(\mathbf{g}_2^i\right)_k$, and one should count the number of occurrences that this difference is positive. For this, we use the Heaviside step function (see Section 8.1.3), and study

$$\left(\mathbf{n}_{1>2}^i\right)_k = \Theta\left(\left(\mathbf{g}_1^i\right)_k - \left(\mathbf{g}_2^i\right)_k\right) \tag{8.89}$$

which is 1 if $\left(\mathbf{g}_1^i\right)_k$ is larger than $\left(\mathbf{g}_2^i\right)_k$, and 0 otherwise. Aggregation over all Monte Carlo trials yields

$$\left(\mathbf{n}_{1>2}\right)_k = \sum_{i=1}^{N} \left(\mathbf{n}_{1>2}^i\right)_k \tag{8.90}$$

We can now study quantities such as $\left(\mathbf{n}_{1>2}\right)_k$, $\left(\mathbf{n}_{2>1}\right)_k$, $\left(\mathbf{n}_{1>3}\right)_k$, $\left(\mathbf{n}_{3>1}\right)_k$, $\left(\mathbf{n}_{2>3}\right)_k$, $\left(\mathbf{n}_{3>2}\right)_k$, and arrange these in a tableau as in Table 7.2.

Obviously, we may define this counting procedure also at the level of characterisation:

$$\left(\mathbf{n}_{1>2}\right)_k = \sum_{i=1}^{N} \Theta\left(\left(\mathbf{h}_1^i\right)_k - \left(\mathbf{h}_2^i\right)_k\right) \tag{8.91}$$

at the level of normalisation

$$\left(\mathbf{n}_{1>2}\right)_k = \sum_{i=1}^{N} \Theta\left(\left(\tilde{\mathbf{h}}_1^i\right)_k - \left(\tilde{\mathbf{h}}_2^i\right)_k\right) \tag{8.92}$$

Table 8.3: Arrangement of the counting of Monte Carlo trials in which one product alternative has a higher score on a certain environmental aspect k than another product alternative.

	product alternative		
product alternative	1	2	3
1	–	$(\mathbf{n}_{1>2})_k$	$(\mathbf{n}_{1>3})_k$
2	$(\mathbf{n}_{2>1})_k$	–	$(\mathbf{n}_{2>3})_k$
3	$(\mathbf{n}_{3>1})_k$	$(\mathbf{n}_{3>2})_k$	–

or at the level of weighting

$$\mathbf{n}_{1>2} = \sum_{i=1}^{N} \Theta \left(W_1^i - W_2^i \right) \tag{8.93}$$

The counters $\mathbf{n}_{1>2}$ are counters in the sense that they count the number of trials that a certain condition is fulfilled. It seems obvious to derive the fraction of trials

$$\mathbf{f}_{1>2} = \frac{\mathbf{n}_{1>2}}{N} \tag{8.94}$$

where N is the number of Monte Carlo trials. A usual significance criterion of 0.95 can be tested. Thus, if $(\mathbf{f}_{1>2})_k > 0.95$ for a certain environmental aspect k, it means that product alternative 1 has a higher score than product alternative 2 in at least 95% of the Monte Carlo realisations. A convenient interpretation is that the products are said to be significantly different, *i.e.* they are discernible.

A final remark is that $(\mathbf{f}_{1>2})_k$ and $(\mathbf{f}_{2>1})_k$ do not necessarily add to 1. It may happen that for certain Monte Carlo trials $(\mathbf{g}_1)_k$ is equal to $(\mathbf{g}_2)_k$ in which case the two fractions will add up to less than 1. In order to create the tableau of Table 7.2 for m product alternatives, one needs to keep track of $m(m-1)$ counters. This is far less than the requirements in constructing $m-1$ frequency distributions of comparison indices.

Chapter 9

Further extensions*

So far, the theory has been based on a number of simplifying assumptions:

- technologies are linear;

- the analysis is based on the steady-state situation;

- there is no spatial differentiation of interventions or impacts.

This chapter will discuss some ideas with respect to dropping these assumptions. Let it be clear that a full discussion is outside the scope of this book.

9.1 Non-linear models

Recall from Chapter 2, and in particular from Axiom 1 that unit processes are conceived as representing a linear technology. When producing 10 kWh of electricity is associated with a fuel demand of 2 litre, an emission of 1 kg CO_2 and an emission of 0.1 kg SO_2, producing 20 kWh of electricity is associated with a fuel demand of 4 litre, an emission of 2 kg CO_2 and an emission of 0.2 kg SO_2. Such linear technologies can be described by a vector equation

$$\mathbf{p}' = s\mathbf{p} \tag{9.1}$$

where $\mathbf{p}$ denotes the process specification, s the scaling factor for the process, and $\mathbf{p}'$ the process characteristics for the actual production volume involved in the analysis. Written down for the case of four flows, for instance, kWh of electricity, litre of fuel, kg of CO_2, and kg of SO_2, we find

$$
\begin{pmatrix} p_1' \\ p_2' \\ p_3' \\ p_4' \end{pmatrix} = s \begin{pmatrix} p_1 \\ p_2 \\ p_3 \\ p_4 \end{pmatrix} = s \begin{pmatrix} -2 \\ 10 \\ 1 \\ 0.1 \end{pmatrix}
\tag{9.2}
$$

Assuming the first flow, representing kWh of electricity, to be the independent variable, and using the other three flows as the dependent variables, we can eliminate the scaling factor s from this equation to find

$$
\begin{pmatrix} p_2' \\ p_3' \\ p_4' \end{pmatrix} = \frac{p_1'}{p_1} \begin{pmatrix} p_2 \\ p_3 \\ p_4 \end{pmatrix}
\tag{9.3}
$$

Now as a first step towards generalisation to include non-linear technologies, we may write this as

$$
\begin{pmatrix} p_2' \\ p_3' \\ p_4' \end{pmatrix} = \mathbf{f}_1 \left(p_1', p_1, \begin{pmatrix} p_2 & p_3 & p_4 \end{pmatrix}^{\mathrm{T}} \right)
\tag{9.4}
$$

where $\mathbf{f}_1(\cdot)$ is a vector-valued function. This which may also be written as

$$
\begin{pmatrix} p_2' \\ p_3' \\ p_4' \end{pmatrix} = \mathbf{f}_1 \left(p_1', \begin{pmatrix} p_1 & p_2 & p_3 & p_4 \end{pmatrix}^{\mathrm{T}} \right) = \mathbf{f}_1 \left(p_1', \mathbf{p} \right)
\tag{9.5}
$$

In the present case, the function $\mathbf{f}_1(\cdot)$ is a linear one:

$$
\mathbf{f}_1 \left(p_1', \mathbf{p} \right) = \frac{p_1'}{p_1} \begin{pmatrix} p_2 \\ p_3 \\ p_4 \end{pmatrix}
\tag{9.6}
$$

The subscript 1 in $\mathbf{f}_1$ serves to indicate that this is the production function for flow 1, in this case kWh of electricity.

As a second step, we can now insert non-linear functions for $\mathbf{f}_1(\cdot)$. These non-linear functions may take process technological relationships and economies of scale into account. We refer to the literature on economic equilibrium models (*e.g.*, Dinwiddy & Teal, 1988) and on process technology models (*e.g.*, Schuler, 1995) for more information. An interesting elaboration of the incorporation of non-linear relationships, accounting for economic behaviour, is provided by Kandelaars (1999).

Of course, there is a problem in specifying these production function. In practice, it is already difficult to find a sufficient amount of process data of sufficient quality for the linear case. Estimating non-linear production functions means that functional relationships must be postulated on the basis of empirical observations and theoretical considerations, deriving from thermodynamics, microeconomics, and probably some more disciplines, and that the coefficients that show up in these functional relationships must be calibrated on the basis of several observations.

Apart from these problems in finding the functional relationships, *i.e.* in formulating the inventory problem, there is a problem in solving it. Let us partition the previous example in terms of two economic flows (kWh of electricity and litre of fuel) and two environmental flows (kg of CO_2 and kg of SO_2). We then can write the production function for the two economic flows as

$$\begin{pmatrix} a'_2 \\ b'_1 \\ b'_2 \end{pmatrix} = \mathbf{f}_1\left(a'_1, \mathbf{p}\right) \tag{9.7}$$

and

$$\begin{pmatrix} a'_1 \\ b'_1 \\ b'_2 \end{pmatrix} = \mathbf{f}_2\left(a'_2, \mathbf{p}\right) \tag{9.8}$$

These two equations must be solved simultaneously. Systems of equations of the form

$$\begin{cases} a'_1 = f_1(a'_2) \\ a'_2 = f_2(a'_1) \end{cases} \tag{9.9}$$

can be extremely hard to solve, depending on the mathematical details of $f_1(\cdot)$ and $f_2(\cdot)$. For a larger system, with hundreds of economic flows and an equally large number of production functions, the task may become unsolvable in an analytical way. Dedicated algorithms, for instance of the Marquardt-type, may be needed to address such systems (Bevington & Robinson, 1992).

The focus of LCA is system-wide on a microscopic level. This can only be achieved on the expense of ignoring certain details, for instance, non-linearities. Including non-linear relationships is likely to imply a more modest ambition with respect to either the system-wide character or the microscopic character.

Quite another matter is the incorporation of non-linear relationships in impact assessment. Section 8.1 has discussed the use of characterisation factors and weighting factors, and Section 8.1.2 has in particular shown how characterisation factors can be derived from non-linear characterisation models under the assumption of an infinitesimally small change on top of a constant background. Thus, in principle, the use of non-linear characterisation models has already been considered, but only under certain restrictions. We may wish to escape from these limitations, and propose to use non-linear relationships directly instead of linearised ones that are derived from the non-linear ones. For this, we may bring into mind the general relationship given in (8.7):

$$h_i = \eta_i(\mathbf{g}) \tag{9.10}$$

where $\eta_i(\cdot)$ is a non-linear function. Use of such models is straightforward, but one should keep in mind the following pitfalls:

- there are no longer characterisation factors, so each model must be used in each case study separately;

- a contribution analysis that decomposes a result into contributing elements is no longer possible.

What has been discussed here for characterisation naturally applies to weighting as well. Non-linear weighting models of the form

$$W = \omega(\tilde{\mathbf{h}}) \tag{9.11}$$

might replace (8.37). An important reason for doing so is to incorporate non-compensatory aspects. Non-linear models for impact assessment has, as far as we know, not been employed in LCA.

9.2   Spatially differentiated models

Essentially and originally, LCA is not a spatially explicit model. Unit processes that are connected to a product are situated all over the world, even for a product that is labeled to be 'made in Holland.' The origin of this is the fact that electricity, steel, and many other commodities are mined or produced in different countries, and that many of these goods are bought from a world market. Of a tomato bought in a supermarket, it is difficult to point out where it was grown. It is even more difficult to find out where the artificial fertiliser came from, and it is impossible to tell which electric power plants were used to cultivate it. This is a theoretical problem. Much sooner, one will encounter practical problems, related to the unavailability of data at the level of individual farmers or companies or to the confidential nature of these data. For many processes one must rely on national, continental, or global averages. Then, it is questionable to what extent it is useful to specify the locations of emissions for those few processes for which more specific data are available.

From a computational perspective, the incorporation of spatial detail is easy. As we have seen, flows of a different type can be distinguished by giving them a separate entry. In previous examples, we had separate rows for a product of brand X and a product of brand Y. This same strategy can be used to distinguish French electricity from German electricity and Swedish sulphur dioxide from Spanish sulphur dioxide. In the LCA of a TV, used in France and made in Germany, one thus selects French electricity for the use phase and German electricity for the manufacturing stage. These two flows are produced by two different processes, which may have emissions of pollutants to a common (European or global) air, or to French air and German air respectively. Characterisation can proceed by regional differentiation of characterisation factors, for instance when the fate of chemicals in the environment depends on the temperature and the soil composition, or when the effects of chemicals depends on the vegetation types found in those regions. Even though the basic model is intended to be used for a model without regional differentiation, the standard subdivision of the environment into air, water and soil can easily be refined into regions of air, water and soil, where the regions can be defined according to political, climatic, ecological or other criteria. In addition, the compartments air, water and soil can be subdivided into compartments such as indoor air, urban air, rural air, rivers, lakes, seas, sand, clay, natural soil, agricultural soil, and so on.

Even if a high degree of spatial detail is added, the model treats the spatial dimension as a discrete attribute. A process takes place in France or in Germany, or perhaps, when talking about Western-European average steel, for 30% in France and for 40% in Germany. There are ecological models in which the spatial dimension is modeled as a continuous variable, see Krewitt *et al.* (1998) for a discussion of the use of the EcoSense model in LCA. Obviously, the basic model for LCA cannot be changed to accommodate for an infinitely large spatial detail. Some finite grid size must be chosen. An inventory vector $\mathbf{g}(x, y)$ will never be produced by an LCA.

9.3   Dynamic models

A basic principle of the inventory model of Chapter 2 is that, although processes may be specified in the temporal and spatial domain, the final demand vector **f** and the inventory vector **g** are obtained by aggregating all processes within the product system:

$$\forall j : \sum_{\forall i} a_{ji} s_i = f_j \tag{9.12}$$

and

$$\forall k : g_k = \sum_{\forall i} b_{ki} s_i \tag{9.13}$$

respectively. Note especially that the inventory vector **g** is not specified as a pattern in time. A time-dependent vector would, for instance, allow one to distinguish emissions last year ago from those next year, emissions in summer from those in winter, and emissions during working hours from those at night. This would require several adaptations of the LCA-procedure.

   In making LCA a dynamic model, processes must be specified according to the time at which they are active for the product under review. A TV that is bought today may have been assembled two months ago, and the components of which it is made may have been produced one month earlier. That is, typical waiting times in production and consumption form an essential ingredient in constructing a dynamic model. Like we before made a spatial differentiation between French and German electricity, electricity production processes, and air, we can make a temporal differentiation between electricity produced now and two months ago, with possibly different technological characteristics (efficiency or fuel mix may have changed) and atmospheric characteristics (temperature may have changed). The differentiation in time thus works out quite along the lines of the differentiation in space. And again, the extent to which differentiation is feasible will be quite restricted in practice. In any case, time must be treated as a discrete parameter, so that one may distinguish carbon monoxide at 1 January, at 2 January, etc. or at 1 am, at 2 am, etc but never as a continuous variable. We will thus never end up with a vector $\mathbf{g}(t)$.

   It should be noted that dynamic input-output analysis (see Chapter 5 for more information on IOA) is a topic that is more developed than dynamic LCA. See, for instance, Duchin & Szyld (1985) and Perrings (1987) for environmental extensions.

Chapter 10

Issues of implementation*

In this chapter, we discuss a number of topics that relate not so much to the theoretical mathematical aspects of the LCA, but rather to practical implementational aspects. This chapter does not present a complete discussion of algorithms, programming details, or other technical issues. For this, the reader is referred to books like Press *et al.* (1992). Here, we will only address certain specific issues that arise mainly by the special circumstances that are offered by LCA.

10.1   Sparse matrices and location matrices

As noted earlier, matrices in LCA tend to be quite large. A typical technology matrix may consist of 500 to 1000 processes (columns) that are connected by an approximately equal number of economic flows (rows). An intervention matrix has the same number of processes (columns), but sometimes an even larger number of environmental flows (rows), because many chemicals that can be released are included for atmospheric, aquatic and terrestrial emission compartments. When spatial differentiation is introduced, distinguishing between hundred countries or regions, the number of rows will increase substantially. The characterisation matrix contains this same number of environmental flows (columns), but the number of impact categories (rows) is more restricted. A number of ten to twenty impact categories is standard practice. However, when one wishes to include sensitivity analyses, *e.g.*, with global warming potentials for different time horizons (Guinée *et al.*, 2002) and toxicity potentials for different cultural perspectives (Hofstetter, 1998), a number of 50 to 100 impact categories will show up. Anyhow, the three matrices mentioned ($\mathbf{A}$, $\mathbf{B}$ and $\mathbf{Q}$) typically require $10,000$ to $1,000,000$ places. A place typically looks like -3.15×10^5: it requires a signed floating-point representation; see Gentle (1998). In Delphi, which is the programming language for CMLCA, the most widely used floating-point variables are `single`, requiring 4 bytes, `double`, requiring 8 bytes, and `extended`, requiring 10 bytes. Fewer bytes means less significant digits and a more restricted range for the exponent. The data type `single` has 7 to 8 significant digits, and permits values ranging from 1.5×10^{-45} to 3.4×10^{38}. This will suffice for LCA. With a 4-byte representation, one would need in the worst case 3 times $1,000,000$ time 4 byte, which is 12 MB. Then, we need to represent standard deviations of all these data, allocation factors, and many more additional items. All in all, a quite large memory occupation will result.

The matrices mentioned are not only large, they contain many zeros as well. This can be seen from the following facts.

- For one particular unit process, one has typically ten to twenty inputs and outputs of economic flows. For instance, a process like production of steel has inputs like iron, electricity, and equipment, and outputs like steel of various sorts. But it has no inputs or outputs of naphtha, tomatoes, or concrete. This means that the process vector contains a few hundreds of zeros for every process, and the technology matrix contains several thousands of zeros.

- For one particular unit process, one has typically ten to twenty inputs and outputs of environmental flows. For instance, a process like mining of iron ore has inputs like iron ore and outputs like a number of heavy metals to soil and water. But it has no inputs or outputs of whales, 1-butyl acetate, or aldicarb. This means that the process vector contains a few hundreds of zeros for every process, and the intervention matrix contains several thousands of zeros.

- For one particular impact category, one may have a few to many contributing environmental flows. For instance, an impact category like acidification has characterisation factors for sulphur oxides, nitrogen oxides, ammonia and a few other substances, emitted to several compartments. But it has no factors for chromium, aldicarb, or iron ore. This means that the characterisation vector contains a few hundreds of zeros for every impact category, and the characterisation matrix contains several thousands of zeros.

- The respective matrices with standard deviations contain only nonzero information when there is a non-zero position in the technology,

> intervention, or characterisation matrix. And, given the limited availability of uncertainty estimates, it will contain even more zeros.

A matrix with a large proportion of zeros in it is called a sparse matrix.

Notice that, strictly speaking, there is a difference between "known to be zero" and "not specified." If there is no data on 1-butyl acetate emissions in connection to the process of mining of iron ore, we are just ignorant about these emissions. That is different from knowing that there is no emission. There are a number of proposals on how to specify such data items: not with a "0", but with a "–" or a "N/A," for instance. Useful as this may be, it cannot be applied in the computation. One cannot invert a matrix with "–" or "N/A" as elements. In computation, all such items are replaced with "0" in the end.

Large sparse matrices can be conveniently represented using the concept of location matrices. A location matrix is a matrix which can only contain non-negative integers. These integers point to vector addresses that contain the floating-point information on the real magnitude. One still needs a large matrix, but this contains non-negative integers, requiring far less memory than signed floating-point numbers. In Delphi, the data types byte, requiring 1 byte, word, requiring 2 bytes, and longword, requiring 4 bytes, are defined. The values of the data type byte range from 0 to 255, for word to 65, 535, and for longword to 4, 294, 967, 295. For most purposes in LCA, the type byte or word will suffice.

As an example, consider the matrix

$$
\begin{pmatrix}
0 & 0 & -2.3 \times 10^{12} & 0 & 0 \\
-12.6 & 0 & 0 & 0 & 7.4 \times 10^{-6} \\
0 & 0 & 0 & 4.2 \times 10^{5} & 0 \\
0 & 0 & 0 & 0 & 0 \\
-0.12 & 0 & 0 & 193 & 0
\end{pmatrix}
\tag{10.1}
$$

Normally, this requires 25 positions of 4 byte, hence 100 byte. Using location matrices, one would use

$$
\begin{pmatrix}
0 & 0 & 1 & 0 & 0 \\
2 & 0 & 0 & 0 & 3 \\
0 & 0 & 0 & 4 & 0 \\
0 & 0 & 0 & 0 & 0 \\
5 & 0 & 0 & 6 & 0
\end{pmatrix}
\quad \text{and} \quad
\begin{pmatrix}
-2.3 \times 10^{12} \\
-12.6 \\
7.4 \times 10^{-6} \\
4.2 \times 10^{5} \\
-0.12 \\
193
\end{pmatrix}
\tag{10.2}
$$

This requires 25 positions of the type byte (requiring 1 byte), and 6 positions of 4 byte, hence a total of 49 byte. The gain is a factor of

2, but it will approach a factor of 4 for large LCA databases. For instance, a matrix of 1000×1000 containing $5,000$ non-zero numbers, requires $1,000,000 \times 1$ byte $+ 5,000 \times 4$ byte $= 1,020,000$ byte instead of $1,000,000 \times 4$ byte $= 4,000,000$ byte.

10.2   Matrix inversion

There is a large literature available on the implementation of matrix inversion and solving systems of equations; see, *e.g.*, Forsythe & Moler (1967), Jennings & McKeown (1977), Press *et al.* (1992) and Golub & Van Loan (1996) for a discussion. We will not repeat this discussion here. However, we must point out some aspects which are of special interest in connection to LCA.

10.2.1   The inverse

In LCA, inversion is only required for the technology matrix. Important features of the technology matrix are:

- it is quite large, *e.g.*, it may have 1000 rows and columns;

- it is sparse, *i.e.* it contains many zeros;

- it may be ill-conditioned, *i.e.* some elements may be 10^{12} while others are 10^{-12}.

These three facts should be borne in mind when choosing an algorithm for inversion.

First the size. Press *et al.* (1992) discuss that matrix inversion is in general a process of order N^3 or slightly less. This means that inverting a 100×100 matrix requires approximately 1000 times as much time as requiring a 10×10 matrix. Inversion of a large matrix with a few thousands of rows and columns can still require a few seconds to a few minutes on a modern PC. If Monte Carlo simulations are performed (see Section 6.4), the matrix inversion is repeated several hundreds or thousands of times, and a vast computation time is needed. It is therefore important to choose a fast algorithm.

Then the sparsity. Golub & Van Loan (1996, p.133) observe that "algorithms for general matrix problems can be streamlined in the presence of such properties as symmetry, definiteness, and sparsity." The technology matrix of LCA is not symmetric or anti-symmetric, it is not positive or negative definite or semi-definite, but it is just sparse, in an irregular way. Unfortunately, the cases discussed by these authors does not suggest a particular solution for the typical technology matrix.

Finally, the differences in scale. As observed earlier (in Section 6.6), very large or very small numbers give rise to a high condition number, and this creates numerical instabilities and introduces round-off errors. A rescaling of rows and/or columns may yield a matrix with a much smaller condition number, and a much more stable behaviour. Section 6.6 gives also more information on rescaling. One computational trick that may be particularly useful is to standardise every column of the matrix $\mathbf{A}$. Standardisation is something that is applied often in statistical computations. Consider a vector of values $x_1, x_2, \ldots, x_n$. Standardisation transforms every element into

$$x'_i = \frac{x_i - \bar{x}}{s_x} \tag{10.3}$$

where

$$\bar{x} = \frac{1}{n} \sum_{i=1}^{n} x_i \tag{10.4}$$

is the mean value of the values x_i, and

$$s_x = \sqrt{\frac{1}{n-1} \sum_{i=1}^{n} (x_i - \bar{x})^2} \tag{10.5}$$

is their sample standard deviation. The resulting standardised vector has zero mean and unit standard deviation. A matrix with its columns standardised has a small condition number, which approaches 1 for uncorrelated columns.

The two other features discussed, the matrix being large and sparse, to some extent compensate one another. Standard approaches towards inversion include Gauss-Jordan elimination with or without backsubstitution, LU decomposition, and singular value decomposition; see, *e.g.*, Press *et al.* (1992). It is not clear which approach seems preferable for application in LCA. It seems therefore wise to implement several algorithms, and to examine the comparative performance in practice.

10.2.2   The pseudoinverse

The pseudoinverse has been introduced into a form that is in principle accessible to computation:

$$\mathbf{A}^+ = \left(\mathbf{A}^{\mathrm{T}}\mathbf{A}\right)^{-1}\mathbf{A}^{\mathrm{T}} \tag{10.6}$$

Although theoretically correct, this is not a practical and robust approach. Press *et al.* (1992) propose singular value decomposition as a superior method of obtaining a pseudoinverse. The matrix $\mathbf{A}$ (which may have more rows m than columns n) is decomposed into

$$\mathbf{A} = \mathbf{U}\mathbf{diag}(\mathbf{w})\mathbf{V}^{\mathrm{T}} \tag{10.7}$$

where $\mathbf{U}$ is an $m \times n$ column-orthogonal matrix, $\mathbf{V}$ is an $n \times n$ orthogonal matrix, and $\mathbf{diag}(\mathbf{w})$ is an $n \times n$ diagonal matrix with non-negative elements, which are known as the singular values of $\mathbf{A}$. Using singular value decomposition, the pseudoinverse is given by

$$\mathbf{A}^{+} = \mathbf{V}\mathbf{diag}(\mathbf{w})\mathbf{U}^{\mathrm{T}} \tag{10.8}$$

Algorithms for singular value decomposition have been described extensively; see for instance Press *et al.* (1992) and Golub & Van Loan (1996).

10.3 Statistics of very long series

In Section 8.2.4, Monte Carlo experiments are used to generate a number of model realisations. The results are then processed to find the mean, the standard deviation, the smallest and largest value, and so on. Calculation of a mean can proceed by adding successive terms and at the end dividing by the number of terms. For the smallest and largest value, a termwise updating of the statistic can be used as well. For the standard deviation or variance, things are more complicated. For a certain quantity of interest, x, one has a series of N values x_i. The sample variance is given by

$$s^2(x) = \frac{1}{N-1} \sum_{i=1}^{N} (x_i - \bar{x})^2 \tag{10.9}$$

where

$$\bar{x} = \frac{1}{N} \sum_{i=1}^{N} x_i \tag{10.10}$$

The problem here is that one must store all values x_i in order to compute the variance. As N may be chosen very large, say 1 million, this would lead to large memory requirements.

The problem may be solved by rewriting the formula as

$$s^2(x) = \frac{1}{N-1} \left(\sum_{i=1}^{N} (x_i^2) - \frac{1}{N} \left(\sum_{i=1}^{N} x_i \right)^2 \right) \tag{10.11}$$

Thus, two counters are needed: one for $\sum x_i$ and one for $\sum x_i^2$, and the individual values of x_i need not be stored.

10.4 Design of software for LCA

In Section 1.1.2, we stated that a knowledge of the computational structure of LCA is important for the design and implementation of reliable LCA software. But we should acknowledge that the development of software for LCA in its turn greatly increases the researcher's insight into the computational structure of LCA (*cf.* Heijungs & Guinée (1993) and Vigon (1996)).

As stated in Section 1.1.1, this book does not provide source codes for writing software for LCA. However, this topic is close enough to the book's to devote a few pages to some general principles.

As a general structure of software for LCA, it is useful to distinguish three features:

- an input module, representing a user-interface to entering data and model settings;

- a computational module, responsible for carrying out the mathematical rules without user-intervention;

- an output module, representing a user-interface to obtaining results in an understandable and attractive form.

Clearly, much of the material presented in this book belongs to the second category. In fact, it is our hope that the proposals found in this book may stimulate the development of robust matrix-based computational modules. For the other two aspects, the input and output modules, this book provides no guidance. Nevertheless, some space may be devoted to more clearly develop the separation into three modules.

In the development of CMLCA, as a matrix-based software for LCA, the emphasis has been laid upon the computational module. Some form of an input and output module were needed of course to feed the software with data, and to record the resulting numbers. These two modules are far less advanced than the computational module. If one takes a look at other available software for LCA, one may observe that many programs are better at the input and output sides. This has been at the expense of the computational module.

It appears that there are several competing aspects, perhaps phrased in terms of emphasis (user-friendly versus LCA-driven versus matrix-oriented) and software engineer (IT-expert versus LCA-practitioner versus mathematician). It may be that the conflict even exists at the level of the programming principles and language. Old-fashioned imperative languages, like Fortran, are still superior for 'number crunching,' newer relational languages, like Prolog, perform well for maintaining data structures, and GUI-based languages, like Visual Basic may be best for mouse-oriented and graphical interfaces.

There is therefore a choice to be made: do we seek a language that deals 'reasonably well' for all three purposes, and what is that language? And how should the programming team than look like and be organised? Or do we design three different modules, each in their own programming language, that exchange information in a common standard? We think that this second line may be a promising way to combine the strength of different types of programming languages. The interaction might perhaps proceed smoothly by sharing pieces of programs, for instance using Windows' DLL-structure.

For the input module, one should be aware of several complicating facts. Steen *et al.* (1995, p.9) mention that a 'relational database technology must be the obvious choice' in the case of LCA. They also propose a detailed description of the structure of a database management system for LCA; see Carlson *et al.* (1998) for an overview, and Boustead (1993) for some additional points of interest. Useful as this may be, one should be aware of the fact that a relational database is in no sense matrix-oriented. But a relational database provides a useful step in building matrices. Every process, flow, impact category, etc. can be addressed by reference to its number in the table in which it is stored.

In the general computational structure, we have defined a process in terms of the flows that enter and leave it. For instance, we could speak of process number 1 that has an input of 2 units of flow number 1 and an output of 10 units of flow number 2. In principle, this representation can be maintained for the input module as well. But for reasons of connection to a user, we must define the name of the process, the name of the flows that enter and leave it, and the name of the units in which these flows are specified. Thus, one then speaks of the process with the name production of electricity, that has an input of 2 litre of fuel and an output of 10 kWh of electricity. Specification of additional attributes is often appropriate, for instance an indication of the location of the process, the time at which the process operates, the data sources, an indication of the uncertainties, and so on. Recall, moreover, that we need at some stage of the calculation that flows are distinguished in a set of economic flows and a set of environmental flows, and that the economic flows are to be divided further into a set of goods and a set of wastes. Such partitioning may be made in the computational module, but one can say that it is likely that such information is already known at the stage of data collection, so that it is appropriate to specify this as a part of the database information, as an attribute of each flow.

Besides the computational items, there are many problems related to procedural aspects in the input module. Data is often shared between several persons or institutes, and all sorts of rules should prevent unclarities on who is allowed to add, change or delete data items. Moreover, a database will often consist of general available data (like those of Frischknecht *et al.* (1993)) and specific user-collected data. When part of the general available data reappears in an updated form, care should be taken that only part of the data is removed. It will probably be necessary to establish a unique coding system for processes, economic flows, environmental flows, and impact categories. For environmental flows, the CAS-numbers are often used, but these provide no sufficient solution: an indication of the compartment (air, soil, etc.) must be added, substance groups (PAH, VOC, etc.) are not covered, and certain substances that appear in more than one form and that behave environmentally different (like Cr^{3+} and Cr^{6+}) cannot be distinguished. For processes and economic flows, several national statistics provide uniform nomenclature and coding at a macro-level. Distinction at a micro-level is often only possible at a company-internal way. Clearly, this issue needs to be addressed in a practical implementation, such as ECOINVENT 2000 (Frischknecht, 2001).

Another interesting issue emerges at the interface of database management and computation. It has to do with the problem of selection and matching of processes and flows on the one hand and of flows and impact categories on the other hand. It may be the case that a process database contains several descriptions of steel production processes: according to the technology of factory X in the year 2000, of factory Y in the year 2001, of the average Western-European average in the last decade, etc. Not all these process specification need to be positioned in the process matrix. Hence the need of a selection step. Next, the chosen process or processes may produce a specified flow, for instance steel of brand X in the year 2000, of brand Y in the year 2001, etc. whereas a certain production process is specified to

require steel of brand Z in the year 1999. It will then be needed to match
different flows to one another. Effectively, this comes down to establishing
the equivalence of steel of brand X in the year 2000 and steel of brand Z
in the year 1999. The computational details of this matching are described
in Section 3.4.4.

For the output module, there is a wide range of possibilities. Tabu-
lar results may be given at the levels of scaling factors, inventory tables,
characterisation, normalisation, and weighting. This may be done for a
single products or for more than one product alternative simultaneously.
Graphical results may be derived from such tables, especially for comparing product alternatives and along with a contribution analysis, in which
the share of processes and flows in a certain results are indicated. Statistical information may visualised using confidence intervals or frequency
distributions. See also Heijungs & Kleijn (2001) for a number of possibilities and pitfalls. For a proper functioning of an output module, an
indication of names of processes and flows, and of the units in which they
are expressed, is required. While the computational module operates at
an abstract level, the available information on naming and units must be
transferred from the input module to the output module. It may sometimes
be convenient to allow for quick jumping from a contribution analysis to
the process specification of dominant process. This requires a quite strong
connection between the output module and the input module.

Figure 10.1 shows the general structure of an LCA-program designed
according to the principles outlined above, which is at present only an ideal.

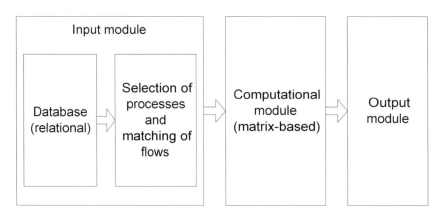

Figure 10.1: General design of LCA-software, distinguishing an input module, a computational module and an output module.

Appendix A

Matrix algebra

This appendix gives a brief summary of those elements of matrix algebra that are needed in this book. It is not intended as a way to learn matrix algebra. Rather, it aims to be helpful to readers who have studied the subject, but who have forgotten the details.

A.1   General concepts

A matrix is nothing more than a collection of numbers arranged in a rectangular grid. For convenience, the grid is normally delimited with parenthesis or brackets. Matrices are conventionally indicated with a bold capital, like $\mathbf{X}$ or $\mathbf{\Lambda}$. An example of a matrix is $\begin{pmatrix} 1 & 0 & 0.5 \\ 0 & -3 & 2 \end{pmatrix}$. The numbers in a matrix are referred to as its elements. Individual elements can be referred to by adding subscripts to the symbol that indicates the matrix itself. The convention is to use the first index for the row, and the second index for the column. Furthermore, counting proceeds from top to bottom for rows and from left to right for columns. For instance, the symbol $(\mathbf{X})_{13}$ has the value 0.5 in the above example. In a larger matrix, the notation $(\mathbf{X})_{123}$ is ambiguous, and a comma will therefore be used to separate the index for row and that for columns, distinguishing $(\mathbf{X})_{1,23}$ and $(\mathbf{X})_{12,3}$. The number of rows and the number of columns define the dimension of a matrix. The example matrix is of dimension 2×3. Obviously, the number of columns or rows cannot be smaller than 1. The elements that form the matrix can be of a predefined set. Throughout this book, the tacit assumption is that they are in the set of real numbers $\mathbb{R}$.

A matrix with 1 column is sometimes referred to as a column vector, or vector in short. The notation for vectors employs a bold lowercase letter, *e.g.*, $\mathbf{x}$. For vectors, it suffices to specify the number of rows. Hence a matrix of dimension 2×1 is the same as a vector of length 2. A matrix with only one row is referred to as a row vector. Although row vectors are important in certain fields of study, they will be rarely used in this book. When they are used, they will be written as a transposed column vector, *e.g.*, $\mathbf{x}^{\mathrm{T}}$. The idea of transposition is discussed in more detail below, in Section A.3.

A matrix of dimension 1×1 is seen a scalar. An italicised symbol (uppercase or lowercase) is used for it, *e.g.*, X. Because the individual elements of a matrix are scalars, the element $(\mathbf{X})_{12}$ can also be referred to as x_{12}. A general form of a matrix $\mathbf{X}$ of dimension $m \times n$ is thus

$$
\mathbf{X} = \begin{pmatrix} x_{11} & x_{12} & \cdots & x_{1n} \\ x_{21} & x_{22} & \cdots & x_{2n} \\ \cdots & \cdots & \cdots & \cdots \\ x_{m1} & x_{m2} & \cdots & x_{mn} \end{pmatrix} \tag{A.1}
$$

When $m = n$, the matrix is said to be square. Many properties and operations on matrices that are discussed below, like the determinant and inversion, are only defined for square matrices. When $m \neq n$, the matrix is said to be rectangular. Observe that this can imply two forms of rectangularity: $m > n$ or $m < n$. In the book, we will be concerned mostly with square matrices or rectangular matrices having more rows than columns.

A.2   Special matrices

Certain matrices occur so often that special names and symbols have been given to them. These include the null matrix $\mathbf{0}$, which consists of zero elements only and the identity matrix $\mathbf{I}$, which is normally square and has zeros everywhere except for the diagonal that contains ones. It is usual to indicate the dimension of these matrices with subscripts. For instance, $\mathbf{0}_{23}$ indicates a null matrix with two rows and three columns, and $\mathbf{I}_3$ an identity matrix with three rows and three columns. In many cases, however, it is not necessary to add these subscripts, as the dimension will be clear from the context. A null vector is a vector that consists of zeros only. Notice that we may write both the null vector and the null matrix as $\mathbf{0}$. This will rarely lead to ambiguities. A vector consisting of ones only is usually indicated as $\mathbf{1}$.

A matrix $\mathbf{X}$ is called positive definite when all elements $x_{ij} > 0$, in which case one writes $\mathbf{X} > 0$, and positive semi-definite when all elements

$x_{ij} \geq 0$, for which one writes $\mathbf{X} \geq 0$. The definition of negative definite and semi-definite matrices is analogous.

A square matrix $\mathbf{X}$ with $x_{ij} = x_{ji}$ for all i and j is called symmetric. If $x_{ij} = -x_{ji}$ (and hence $x_{ii} = 0$), it is called anti-symmetric.

A.3   Basic operations

The well-known operations of addition, multiplication, division and so on are as yet undefined for matrices. The following definitions are convention-ally adopted.

Two matrices $\mathbf{X}$ and $\mathbf{Y}$ of the same dimension can be added by element-wise addition:

$$(\mathbf{X} + \mathbf{Y})_{ij} = \mathbf{X}_{ij} + \mathbf{Y}_{ij} \tag{A.2}$$

Addition of a matrices with unequal dimension is not defined. This applies in particular to addition of a matrix and a scalar.

Any matrix can be multiplied with a scalar by multiplication of each element with that scalar:

$$(c\mathbf{X})_{ij} = c\mathbf{X}_{ij} \tag{A.3}$$

A special case is when $c = -1$. The resulting matrix is written as $-\mathbf{X}$. Subtracting two matrices $\mathbf{X}$ and $\mathbf{Y}$ of equal dimension can then be regarded as adding matrices $\mathbf{X}$ and $-\mathbf{Y}$.

Two matrices may also be multiplied. In fact, various such products are defined. The most common one is referred to as the matrix product, or Cauley product. The matrix product $\mathbf{XY}$ is only defined when the number of columns of $\mathbf{X}$ is equal to the number of rows of $\mathbf{Y}$. In this case the matrices are said to be conformable. Matrix multiplication involves a sum over all the columns of the first matrix, or for that sake, all the rows of the second matrix:

$$(\mathbf{XY})_{ij} = \sum_{k=1}^{m} \mathbf{X}_{ik}\mathbf{Y}_{kj} \tag{A.4}$$

The matrix product of a matrix $\mathbf{X}$ of dimension $l \times m$ and a matrix $\mathbf{Y}$ of dimension $m \times n$ is itself a matrix of dimension $l \times n$. When l is not equal to n, the possibility of forming the matrix product $\mathbf{XY}$ implies that the matrix product $\mathbf{YX}$ cannot be formed. But even in the case of square matrices, when $\mathbf{YX}$ can actually be formed, the two matrix products $\mathbf{XY}$ and $\mathbf{YX}$ are in general not equal. Matrix multiplication is, in contrast to ordinary multiplication, not commutative. It is, however, associative: $\mathbf{X}(\mathbf{YZ})$ is equal to $(\mathbf{XY})\mathbf{Z}$ and one may thus write $\mathbf{XYZ}$.

A special case occurs when $n = 1$, *i.e.* when $\mathbf{Y}$ is a vector $\mathbf{y}$. The product of a matrix and a conformable vector is

$$(\mathbf{Xy})_i = \sum_{k=1}^{m} \mathbf{X}_{ik}\mathbf{y}_k \tag{A.5}$$

It is thus a vector, with a possibly different number of rows than $\mathbf{y}$.

Straightforward application of the rules gives the following special cases:

$$\mathbf{IX} = \mathbf{XI} = \mathbf{X} \tag{A.6}$$

and

$$\mathbf{0X} = \mathbf{X0} = \mathbf{0} \tag{A.7}$$

for any matrix $\mathbf{X}$.

There is no such thing as dividing two matrices. There is, however, an operation called inversion that bears some resemblance to the inversion of ordinary numbers. For any number x, except 0, we may form a number y such that the product of y and x is equal to 1. This number y is referred to as the inverse of x, and we write it as $1/x$ or as x^{-1}. The generalisation to matrices is as follows. When a square matrix $\mathbf{Y}$ exists, such that the matrix product of $\mathbf{Y}$ and the square matrix $\mathbf{X}$ is equal to the identity matrix $\mathbf{I}$, the matrix $\mathbf{Y}$ is said to be the inverse of matrix $\mathbf{X}$. By analogy, one writes $\mathbf{X}^{-1}$ for the inverse of $\mathbf{X}$. The generalisation of the condition that a scalar x must be different from 0 is that the determinant (see below in Section A.4) of a matrix $\mathbf{X}$ must be different from 0. Every square matrix with nonzero determinant is invertible, and has a unique inverse. Conversely, a square matrix with determinant zero is not invertible. Such a matrix is said to be singular. Otherwise, it is non-singular.

For a square invertible matrix $\mathbf{X}$ and its inverse $\mathbf{Y}$ one has by definition

$$\mathbf{YX} = \mathbf{I} \tag{A.8}$$

A simple consequence is that

$$\mathbf{XYX} = \mathbf{X} \tag{A.9}$$

This relationship, however, holds for more matrices than the inverse. In fact, matrices $\mathbf{Y}$ can be found such that it also holds for matrices that are not square and hence not invertible. Such a matrix $\mathbf{Y}$ is referred to as a generalised inverse of $\mathbf{X}$. In contrast to the normal inverse of a square matrix, this matrix is not unique. With certain additional conditions, a unique matrix can be found. This matrix is called the pseudoinverse or Moore-Penrose inverse, and it is denoted as $\mathbf{X}^+$. When $\mathbf{X}$ has dimension $m \times n$, its pseudoinverse has dimension $n \times m$. And when $\mathbf{X}$ is square and invertible, its pseudoinverse is equal to its normal inverse.

Of a quite different nature is the operation of transposition. Above, it was already stated that a row vector may be transposed into a column vector and *vice versa*. More generally, a matrix $\mathbf{X}$ of dimension $m \times n$ can be transposed into a matrix $\mathbf{X}^{\mathrm{T}}$ of dimension $n \times m$ with elements given by

$$x_{ij}^{\mathrm{T}} = x_{ji} \tag{A.10}$$

Transposition can thus be regarded as reflection in the main diagonal. An alternative way to indicate the transpose of a matrix $\mathbf{X}$ is $\mathbf{X}'$.

A diagonal matrix is a square matrix with zeros everywhere except at the main diagonal. A vector $\mathbf{x}$ of dimension n can be converted into a diagonal matrix $\mathbf{X}$ of dimension $n \times n$ with the **diag** operator:

$$\mathbf{diag}(\mathbf{x}) = \begin{pmatrix} x_1 & 0 & \cdots & 0 \\ 0 & x_2 & \cdots & 0 \\ \cdots & \cdots & \cdots & \cdots \\ 0 & 0 & \cdots & x_n \end{pmatrix} \tag{A.11}$$

achieves this. Alternatively, the notation $\hat{\mathbf{x}}$ is used.

A square matrix of dimension $n \times n$ has n^2 minors. Minors are denoted as $\mathbf{X}_{ij}$ where the subscripts i and j can assume any value between 1 and n. The minor $\mathbf{X}_{ij}$ of a matrix $\mathbf{X}$ is a matrix of dimension $(n-1) \times (n-1)$ which is found by removing the ith row and the jth column of $\mathbf{X}$.

A.4 Basic properties

A number of properties of matrices have been defined. Among these are the determinant, the norm, and the eigenvalues.

The determinant of a square matrix $\mathbf{X}$ is a number that characterises the degree of independency of the rows and columns of that matrix. It can be introduced in many ways. Here, we will use a computational form. The determinant of matrices of dimension 1×1 and 2×2 is

$$\det \begin{pmatrix} x_{11} \end{pmatrix} = x_{11}; \quad \det \begin{pmatrix} x_{11} & x_{12} \\ x_{21} & x_{22} \end{pmatrix} = x_{11}x_{22} - x_{21}x_{12} \tag{A.12}$$

For determinants of larger matrices, use is made of the formula

$$\det(\mathbf{X}) = \sum_j (-1)^{i+j} x_{ij} \det(\mathbf{X}_{ij}) \tag{A.13}$$

where $\mathbf{X}_{ij}$ is one of the minors of $\mathbf{X}$. This formula can be recursively applied if necessary. For instance, the determinant of a 4×4-matrix can be expressed as a sum of determinants of 3×3-matrices, which on their turn can be expressed in terms of a sum of 2×2-matrices. The determinant of a matrix $\mathbf{X}$ is sometimes indicated as $|\mathbf{X}|$.

We already saw that the determinant of a singular matrix $\mathbf{X}$ is 0. This happens, for instance, when one of the columns of $\mathbf{X}$ can be expressed as a linear combination of one or more other columns. In that case, there is a non-trivial solution to $\mathbf{xX} = \mathbf{0}$, and the the column vectors are said to be dependent. The degree of dependency may be larger or smaller: if only one column of a matrix of dimension $n \times n$ may be expressed as a linear combination of the other $n - 1$ ones, we say that the rank of the matrix is $n - 1$. If two columns may be expressed as a linear combination of the other $n - 2$ ones, the rank of the matrix is $n - 2$. The rank of a square matrix of dimension $n \times n$ is at most n, in which case the matrix is said to be of full rank. If not, it is said to be rank-deficient.

The product of a square matrix $\mathbf{X}$ and a conformable vector $\mathbf{x}$ is a vector of the same dimension as $\mathbf{x}$. There are special vectors $\mathbf{X}$ (excluding the null vector $\mathbf{0}$) for which this product $\mathbf{Xx}$ is a multiple of $\mathbf{x}$, say $\lambda \mathbf{x}$. Such special vectors $\mathbf{x}$ are referred to as the eigenvectors of $\mathbf{X}$, and the scalars λ are called the associated eigenvalues. Thus, the equation

$$\mathbf{Xx} = \lambda \mathbf{x} \tag{A.14}$$

defines the eigensystem of the matrix $\mathbf{X}$. This may be written as

$$(\mathbf{X} - \lambda \mathbf{I}) \, \mathbf{x} = \mathbf{0} \tag{A.15}$$

and this system of linear equations has a non-trivial solution if and only if

$$\det (\mathbf{X} - \lambda \mathbf{I}) = 0 \tag{A.16}$$

can be solved for $\mathbf{X}$ and λ. This equation is called the characteristic equation. It is a polynomial, which can be solved for λ to yield the eigenvalues of $\mathbf{X}$. When the dimension of $\mathbf{X}$ is $n \times n$, there may be up to n eigenvalues, but some of these may be equal to one another.

Vector norms can be defined as

$$\|\mathbf{x}\|_p = \left(\sum_i |x_i|^p \right)^{1/p} \tag{A.17}$$

where $\mathbf{x}$ represents an arbitrary vector, and p defines the type of metric considered. The norm defined in this way is referred to as the L_p norm.

We will restrict the discussion to the case $p = 2$, in which case the norm is called the Euclidean norm. The subscript for $p = 2$ is then dropped:

$$\|\mathbf{x}\| = \|\mathbf{x}\|_2 = \sqrt{\sum_i |x_i|^2} = \sqrt{\mathbf{x}^T\mathbf{x}} \qquad (A.18)$$

This norm of a vector, and in particular the L_2-norm is sometimes referred to as the length of that vector, and it corresponds to the familiar Pythagorean concept of length. Notice, however, that the term length in some texts may refer to the number of rows of a vector. A vector norm satisfies the triangle inequality

$$\|\mathbf{x} + \mathbf{y}\| \leq \|\mathbf{x}\| + \|\mathbf{y}\| \qquad (A.19)$$

The norm of a matrix is a quantity that is induced by a vector norm. The definition we shall use is

$$\|\mathbf{X}\| = \max_{\|\mathbf{x}\|=1} \|\mathbf{X}\mathbf{x}\| \qquad (A.20)$$

which can be interpreted as the largest possible vector norm of the product $\mathbf{X}\mathbf{x}$ under the condition that the vector norm of $\mathbf{x}$ itself is 1. In some sense, it is thus a measure of the maximum magnification of $\mathbf{X}$. The norm of a matrix is not easy to calculate, as it is involves a maximisation over an infinite number of vectors $\mathbf{x}$. For a square matrix $\mathbf{X}$, things become easier. It is equal to the square root of largest eigenvalue $\lambda_{\max}$ of the product $\mathbf{X}^T\mathbf{X}$:

$$\|\mathbf{X}\| = \sqrt{\lambda_{\max}\left(\mathbf{X}^T\mathbf{X}\right)} \qquad (A.21)$$

Eigenvalues are much easier to calculate. The largest eigenvalue is also known as the spectral radius. In analogy with the triangle inequality for vector norms, there is an inequality for matrix norms which states that for any two conformable matrices $\mathbf{X}$ and $\mathbf{Y}$

$$\|\mathbf{X}\mathbf{Y}\| \leq \|\mathbf{X}\| + \|\mathbf{Y}\| \qquad (A.22)$$

Another norm, that is sometimes employed for its ease in analytical formulas, is the Frobenius norm. It can be found by

$$\|\mathbf{X}\|_F = \sqrt{\operatorname{tr}\left(\mathbf{X}^T\mathbf{X}\right)} \qquad (A.23)$$

where the trace $\operatorname{tr}(\cdot)$ of a square matrix is defined as the sum of its diagonal elements:

$$\sqrt{\operatorname{tr}(\mathbf{X})} = \sum_i x_{ii} \qquad (A.24)$$

The norm of a square invertible matrix $\mathbf{X}$ can be multiplied with the norm of its inverse. The result is referred to as the condition number with respect to inversion, or simply the condition number. It is given the symbol $\kappa(\mathbf{X})$. Thus

$$\kappa(\mathbf{X}) = \|\mathbf{X}\| \, \|\mathbf{X}^{-1}\| \tag{A.25}$$

Notice that, as we have restricted the discussion to Euclidean norms, the condition numbers is also discussed here for Euclidean norms. For other norms, *i.e.* for other values for p than 2, a corresponding condition number can be found. One therefore sees that the condition number is said to be the condition number with respect to a certain norm. In general, these various condition numbers are of the same order of magnitude for a given matrix. Also notice that the inverse matrix of $\mathbf{X}$ is needed in the definition of the condition number. This means that a condition number is only defined for square invertible matrices. It can be shown that the (Euclidean) condition number of a matrix $\mathbf{X}$ is equal to the ratio of the largest and the smallest eigenvalue of that matrix:

$$\kappa(\mathbf{X}) = \frac{|\lambda_{\max}(\mathbf{X})|}{|\lambda_{\min}(\mathbf{X})|} \tag{A.26}$$

This property then serves to generalise the condition number for rectangular matrices and for square but non-invertible (*i.e.* singular) matrices.

A.5 Partitioned matrices

Similar to the construction of a matrix by arranging numbers in a rectangular grid, we may arrange matrices in such a grid to construct a larger matrix. Or, the other way around, we may divide a matrix into a number of smaller matrices, the so-called submatrices. A matrix of the form

$$\mathbf{X} = \begin{pmatrix} \mathbf{X}_{11} & \mathbf{X}_{12} \\ \mathbf{X}_{21} & \mathbf{X}_{22} \end{pmatrix} \tag{A.27}$$

is referred to as a partitioned matrix. Such a matrix is only possible when the dimensions of the various submatrices are conformable. Here, this means that the dimension of $\mathbf{X}_{11}$ is $m_1 \times n_1$, that $\mathbf{X}_{12}$ is $m_1 \times n_2$, $\mathbf{X}_{21}$ is $m_2 \times n_1$ and $\mathbf{X}_{22}$ is $m_2 \times n_2$. Then $\mathbf{X}$ is of dimension $m_1 + m_2 \times n_1 + n_2$. Sometimes, we will indicate the submatrices more explicitly by drawing horizontal and/or vertical lines:

$$\mathbf{X} = \left(\begin{array}{c|c} \mathbf{X}_{11} & \mathbf{X}_{12} \\ \hline \mathbf{X}_{21} & \mathbf{X}_{22} \end{array} \right) \tag{A.28}$$

Most rules for manipulating partitioned matrices are straightforward, for instance, those for addition

$$\begin{pmatrix} \mathbf{X}_{11} & \mathbf{X}_{12} \\ \mathbf{X}_{21} & \mathbf{X}_{22} \end{pmatrix} + \begin{pmatrix} \mathbf{Y}_{11} & \mathbf{Y}_{12} \\ \mathbf{Y}_{21} & \mathbf{Y}_{22} \end{pmatrix} = \begin{pmatrix} \mathbf{X}_{11} + \mathbf{Y}_{11} & \mathbf{X}_{12} + \mathbf{Y}_{12} \\ \mathbf{X}_{21} + \mathbf{Y}_{21} & \mathbf{X}_{22} + \mathbf{Y}_{22} \end{pmatrix} \quad \text{(A.29)}$$

for multiplication

$$\begin{pmatrix} \mathbf{X}_{11} & \mathbf{X}_{12} \\ \mathbf{X}_{21} & \mathbf{X}_{22} \end{pmatrix} \begin{pmatrix} \mathbf{Y}_{11} & \mathbf{Y}_{12} \\ \mathbf{Y}_{21} & \mathbf{Y}_{22} \end{pmatrix} = \\ \begin{pmatrix} \mathbf{X}_{11}\mathbf{Y}_{11} + \mathbf{X}_{12}\mathbf{Y}_{21} & \mathbf{X}_{11}\mathbf{Y}_{12} + \mathbf{X}_{12}\mathbf{Y}_{22} \\ \mathbf{X}_{21}\mathbf{Y}_{11} + \mathbf{X}_{22}\mathbf{Y}_{21} & \mathbf{X}_{21}\mathbf{Y}_{12} + \mathbf{X}_{22}\mathbf{Y}_{22} \end{pmatrix} \quad \text{(A.30)}$$

and for transposition

$$\begin{pmatrix} \mathbf{X}_{11} & \mathbf{X}_{12} \\ \mathbf{X}_{21} & \mathbf{X}_{22} \end{pmatrix}^{\mathrm{T}} = \begin{pmatrix} \mathbf{X}_{11}^{\mathrm{T}} & \mathbf{X}_{21}^{\mathrm{T}} \\ \mathbf{X}_{12}^{\mathrm{T}} & \mathbf{X}_{22}^{\mathrm{T}} \end{pmatrix} \quad \text{(A.31)}$$

For inversion, a more complicated formula can be derived (Harville, (1997, p.99)):

$$\begin{pmatrix} \mathbf{X}_{11} & \mathbf{X}_{12} \\ \mathbf{X}_{21} & \mathbf{X}_{22} \end{pmatrix}^{-1} = \begin{pmatrix} \mathbf{X}_{11}^{-1} & \mathbf{0} \\ \mathbf{0} & \mathbf{0} \end{pmatrix} + \\ \begin{pmatrix} -\mathbf{X}_{11}^{-1}\mathbf{X}_{12} \\ \mathbf{I} \end{pmatrix} (\mathbf{X}_{22} - \mathbf{X}_{21}\mathbf{X}_{11}^{-1}\mathbf{X}_{12})^{-1} \begin{pmatrix} -\mathbf{X}_{21}\mathbf{X}_{11}^{-1} & \mathbf{I} \end{pmatrix} \quad \text{(A.32)}$$

where $\mathbf{X}_{11}$ must be square and non-singular, $\mathbf{X}_{22}$ must be square, and $\mathbf{X}_{22} - \mathbf{X}_{21}\mathbf{X}_{11}^{-1}\mathbf{X}_{12}$ must be non-singular. An important special case of this equation is the one that the off-diagonal submatrices are $\mathbf{0}$, in which case the expression reduces to

$$\begin{pmatrix} \mathbf{X}_{11} & \mathbf{0} \\ \mathbf{0} & \mathbf{X}_{22} \end{pmatrix}^{-1} = \begin{pmatrix} \mathbf{X}_{11}^{-1} & \mathbf{0} \\ \mathbf{0} & \mathbf{X}_{22}^{-1} \end{pmatrix} \quad \text{(A.33)}$$

A.6 Systems of linear equations

A system of linear equations is a set of set of m equations in n unknowns $x_1, x_2, \ldots, x_n$:

$$\begin{cases} a_{11}x_1 + a_{12}x_2 + \cdots + a_{1n}x_n = y_1 \\ a_{21}x_1 + a_{22}x_2 + \cdots + a_{2n}x_n = y_2 \\ \qquad \cdots \\ a_{m1}x_1 + a_{m2}x_2 + \cdots + a_{mn}x_n = y_m \end{cases} \quad \text{(A.34)}$$

where $a_{11}, \ldots, a_{mn}$ and $y_1, \ldots, y_m$ are fixed and given. This can be written in matrix form as

$$\mathbf{A}\mathbf{x} = \mathbf{y} \tag{A.35}$$

The problem is to solve the system of equations for the unknown vector $\mathbf{x}$, given $\mathbf{A}$ and $\mathbf{y}$.

We first discuss the case of $m = n$, *i.e.* when $\mathbf{A}$ is square. In this case, $\mathbf{A}$ may be invertible. If it is so, left and right side of the matrix equation can be multiplied with $\mathbf{A}^{-1}$ to yield

$$\mathbf{A}^{-1}\mathbf{A}\mathbf{x} = \mathbf{A}^{-1}\mathbf{y} \tag{A.36}$$

Because the product of a square matrix and its inverse is equal to the identity matrix, this immediately leads to an explicit expression for $\mathbf{x}$:

$$\mathbf{x} = \mathbf{A}^{-1}\mathbf{y} \tag{A.37}$$

When, however, $\mathbf{A}$ is not invertible, the system of linear equations can not be solved.

Next, we discuss the case of an overdetermined system of equations with $m > n$, *i.e.* when there are more equations than unknowns. In that case, we can multiply left and right side with $\mathbf{A}^{T}$, to obtain

$$\mathbf{A}^{T}\mathbf{A}\mathbf{x} = \mathbf{A}^{T}\mathbf{y} \tag{A.38}$$

The product of a matrix and its transpose is square, and it may be invertible. If it is so, we can form

$$\left(\mathbf{A}^{T}\mathbf{A}\right)^{-1}\mathbf{A}^{T}\mathbf{A}\mathbf{x} = \left(\mathbf{A}^{T}\mathbf{A}\right)^{-1}\mathbf{A}^{T}\mathbf{y} \tag{A.39}$$

which reduces to

$$\mathbf{x} = \left(\mathbf{A}^{T}\mathbf{A}\right)^{-1}\mathbf{A}^{T}\mathbf{y} \tag{A.40}$$

which is again an explicit solution to $\mathbf{x}$. When $\mathbf{A}^{T}\mathbf{A}$ is not invertible, the system of equations can again not be solved.

Notice that there is a subtle difference. In the square case, $\mathbf{A}^{-1}\mathbf{A}\mathbf{x} = \mathbf{A}^{-1}\mathbf{y}$ implies that $\mathbf{A}^{-1}(\mathbf{A}\mathbf{x} - \mathbf{y}) = \mathbf{0}$ which must mean that indeed $\mathbf{A}\mathbf{x} = \mathbf{y}$. But for the rectangular case, $\mathbf{A}^{T}\mathbf{A}\mathbf{x} = \mathbf{A}^{T}\mathbf{y}$ implies $\mathbf{A}^{T}(\mathbf{A}\mathbf{x} - \mathbf{y}) = \mathbf{0}$ which does not mean that $\mathbf{A}\mathbf{x} = \mathbf{y}$. It only means that the difference

$$\mathbf{e} = \mathbf{A}\mathbf{x} - \mathbf{y} \tag{A.41}$$

has a minimum norm. The solution $\mathbf{x}$ in that case represents the least squares solution to an overdetermined system of equations. Occasionally, the norm of the residual vector $\mathbf{e}$ may be zero, in which case the system of equations is solved exactly.

When, in the square case, $\mathbf{X}$ is not invertible, in the rectangular case, $\mathbf{X}^{\mathrm{T}}\mathbf{X}$ is not invertible, or there are fewer equations than unknown (hence $m < n$), the system of equations is underdetermined. There are infinitely many exact solutions to such a system of equations.

Appendix B

Main terms and symbols

This appendix lists the symbols that occur at several places. Thus occasionally introduced symbols are not included in the table.

Symbol	Name, meaning	Defined in Section
$\mathbf{A}$	technology matrix	2.1
$\mathbf{A}_S$	scaled technology matrix	7.1
$\mathbf{B}$	intervention matrix	2.1
$\mathbf{B}_S$	scaled intervention matrix	7.1
$\mathbf{d}$	discrepancy vector	3.1
$\mathbf{f}$	final demand vector	2.1
$\tilde{\mathbf{f}}$	final supply vector	3.1
ϕ	reference flow	2.1
$\mathbf{g}$	inventory vector	2.1
$\dot{\mathbf{g}}$	reference inventory vector	8.1.5
$\boldsymbol{\Gamma}_k$	perturbation matrix for the kth element of the inventory vector	8.2.3
$\mathbf{h}$	impact vector	8.1.4
$\dot{\mathbf{h}}$	reference impact vector	8.1.5
$\tilde{\mathbf{h}}$	normalised impact vector	8.1.5
$\hat{\mathbf{H}}$	classification matrix	8.1.3
$\boldsymbol{\Lambda}$	intensity matrix	2.2
$\mathbf{p}$	process vector	2.1
$\mathbf{P}$	process matrix	2.1
$\mathbf{q}$	system vector	2.1
$\mathbf{q}$	characterisation vector	8.1.4
$\mathbf{Q}$	characterisation matrix	8.1.4

217

s	scaling vector	2.2
$\boldsymbol{\Sigma}_k$	perturbation matrix for the kth element of the scaling vector	8.2.3
w	weighting vector	8.1.7
W	weighted index	8.1.7

Appendix C

Matlab code for most important algorithms

Matlab provides a powerful software package for dealing with the matrix-based formulae developed in this book. Below, examples of the most important pieces of Matlab code are given. A more elaborate version can be obtained from the website mentioned in the Preface.

The first three statements define the input data: technology matrix **A**, intervention matrix **B** and final demand vector **f** for the simple example that recurs at many places in this book:

```
A=[-2 100; 10 0];
B=[1 10; 0.1 2; 0 -50];
f=[0; 1000];
```

The next series of statements provide the central computational commands, yielding the inverse $\mathbf{A}^{-1}$ of the technology matrix, the scaling vector **s**, and the inventory vector **g**:

```
Ainv=inv(A);
s=Ainv*f;
g=B*s;
```

Some auxiliary vectors and matrices (the intensity matrix $\mathbf{\Lambda}$, the vector of final supply $\tilde{\mathbf{f}}$, the discrepancy vector **d**, the process matrix **P** and the system vector **q**) and the scaled technology and intervention matrix $\mathbf{A_S}$ and $\mathbf{B_S}$ are defined below:

```
Lambda=B*Ainv;
```

```
ftilde=A*s;
d=ftilde-f;
P=[A; B];
q=[f; g];
As=A*diag(s);
Bs=B*diag(s);
```

The following statements calculate various perturbation-theoretic quantities, including the condition number of the technology matrix $\kappa(\mathbf{A})$, the matrices of derivatives $\dfrac{\partial \mathbf{s}_k}{\partial \mathbf{A}}$, $\dfrac{\partial \mathbf{g}_k}{\partial \mathbf{A}}$ and $\dfrac{\partial \mathbf{g}_k}{\partial \mathbf{B}}$ and the perturbation matrices for the scaling vector $\mathbf{\Sigma}_k$ and the for inventory vector $\mathbf{\Gamma}_k(\mathbf{A})$ and $\mathbf{\Gamma}_k(\mathbf{B})$:

```
kappa=cond(A);
for k=1:size(s),
    for i=1:size(A,1),
    for j=1:size(A,2),
        dsdA(i,j,k)=-Ainv(k,i)*s(j);
      end
  end
end;
for k=1:size(g),
    for i=1:size(A,1),
    for j=1:size(A,2),
            dgdA(i,j,k)=-Lambda(k,i)*s(j);
      end
  end
end;
for k=1:size(g),
    for i=1:size(B,1),
      for j=1:size(B,2),
        if i==k
            dgdB(i,j,k)=s(j);
        else
            dgdB(i,j,k)=0;
        end
      end
    end
end;
for k=1:size(s),
    for i=1:size(A,1),
```

```
      for j=1:size(A,2),
          sigmaA(i,j,k)=-A(i,j)/s(k)*Ainv(k,i)*s(j);
      end
   end
end;
for k=1:size(g),
   for i=1:size(A,1),
     for j=1:size(A,2),
         gammaA(i,j,k)=-A(i,j)/g(k)*Lambda(k,i)*s(j);
     end
   end
end;
for k=1:size(g),
   for i=1:size(B,1),
     for j=1:size(B,2),
         if i==k
             gammaB(i,j,k)=B(i,j)/g(k)*s(j);
         else
             gammaB(i,j,k)=0;
         end
     end
   end
end;
```

References

Albert, A. *Regression and the Moore-Penrose pseudoinverse*. Academic Press, New York, 1972.

Anonymous. Life-cycle assessment. In: H.-J. Arpe (Ed.) *Ullmann's Encyclopedia of Industrial Chemistry, 5th Edition, Volume B8*. VCH Verlagsgesellschaft, Weinheim, 1995, pp. 585–600.

Apostol, T.M. *Calculus. Volume II. Multi-variable calculus and linear algebra, with applications to differential equations and probability. Second edition*. John Wiley & Sons, New York, 1969.

Atkinson, K.E. *An introduction to numerical analysis. Second edition*. John Wiley & Sons, New York, 1989.

Ayres, R.U. & A.V. Kneese. Production, consumption, and externalities. *The American Economic Review* LIX (1969), 282–297.

Azapagic, A. & R. Clift. Allocation of environmental burdens by whole-system modelling. The use of linear programming. In: G. Huppes & F. Schneider (Eds.). *Allocation in LCA*. SETAC, Brussels, 1994, pp. 54–60.

Azapagic, A. & R. Clift. Life cycle assessment and multiobjective optimisation. *Journal of Cleaner Production* 7 (1999), 135–143.

Balestra, P. *La dérivation matricielle. Technique et résultats pour économistes*. Sirey, Paris, 1976.

Baumgärtner, S. *Ambivalent joint production and the natural environment. An economic and thermodynamic analysis*. Physica-Verlag, Heidelberg, 2000.

Baumol, W.J. *Economic theory and operations analysis. Third edition*. Prentice-Hall International, Inc., London, 1972.

Berg, N.W. van den, G. Huppes, E.W. Lindeijer, B.L. van der Ven & N.M. Wrisberg. *Quality assessment for LCA*. CML, Leiden, 1999.

Bevington, P.R. & D.K. Robinson. *Data reduction and error analysis for the physical sciences. Second edition*. McGraw-Hill, Inc., New York,

1992.

Boustead, I. & G.F. Hancock. *Handbook of industrial energy analysis.* Ellis Horwood Limited, Chichester, 1979.

Boustead, I. General principles for life cycle assessment databases. *Journal of Cleaner Production* 1:3/4 (1993), 167–172.

Bullard, C.W., P.S. Penner & D.A. Pilati. Net energy analysis. Handbook for combining process and input-output analysis. *Resources and Energy* 1:3 (1978), 276–313.

Carlson, R., A.-M. Tillman, B. Steen & G. Löfgren. LCI data modelling and database design. *International Journal of Life Cycle Assessment* 3:2 (1998), 106–113.

Cheney, W. & D. Kincaid. *Numerical mathematics and computing. Fourth edition.* Brooks/Cole Publishing Company, Pacific Grove, 1999.

Chevalier, J.-L. & J.-F. Le Téno. Life cycle analysis with ill-defined data and its application to building products. *International Journal of Life Cycle Assessment* 1:2 (1996), 90–96.

Chiang, A.C. *Fundamental methods of mathematical economics. Third edition.* McGraw-Hill Book Company, Auckland, 1984.

Ciroth, A. *Fehlerrechnung in Ökobilanzen.* PhD thesis, Technischen Universität Berlin, Berlin, 2001.

Clift, R. R. Frischknecht, G. Huppes, A.-M. Tillman & B. Weidema (Eds.). *Towards a coherent approach to life cycle inventory analysis.* Unpublished manuscript, 1998.

Consoli, F., D. Allen, I. Boustead, J. Fava, W. Franklin, A.A. Jensen, N. de Oude, R. Parrish, R. Perriman, D. Postlethwaite, B. Quay, J. Séguin & B. Vigon. Guidelines for life-cycle assessment. A 'Code of Practice'. Edition I. SETAC, Brussels, 1993.

Copius Peereboom, E., R. Kleijn, S. Lemkowitz & S. Lundie. Influence of inventory data sets on life cycle assessment results. A case study on PVC. *Journal of Industrial Ecology* 2:3 (1999), 109–130.

Coulon, R. V. Camobreco, H. Teulon & J. Besnainou. Data quality and uncertainty in LCA. *Journal of Life Cycle Assessment* 2:3 (1997), 178–182.

Curran, M.A. *Environmental life-cycle assessment.* McGraw-Hill, New York, 1996.

Dijk, N.M. van & K. Sladký. *Perturbation theory for open Leontief input-output models.* Tinbergen Institute, Amsterdam, 1995.

Dinwiddy, C.L. & F.J. Teal. *The two-sector general equilibrium model.* Philip Allan Publishers Limited, Oxford, 1988.

Duchin, F. & A.E. Steenge. Input-output analysis, technology and the environment. In: J.C.J.M. van den Bergh (Ed.). *Handbook of environmental and resource economics.* Edward Elgar, Cheltenham, 1999, pp. 1037–1059.

Duchin, F. & D.B. Szyld. A dynamic input-output model with assured positive output. *Metroeconomica,* 37 (1985), 269–282.

Edey, H.C. & A.T. Peacock. *National income and social accounting. Second revised edition.* Hutchinson University Library, London, 1959.

Engelenburg, B.C.W. van, T.F.M. van Rossum, K. Blok & K. Vringer. Calculating the energy requirements of households purchases. *Energy Policy* 22:8 (1994), 648–656.

Fava, J.A., R. Denison, B. Jones, M.A. Curran, B. Vigon, S. Selke & J. Barnum. *A technical framework for life-cycle assessments.* SETAC, Washington, 1991.

Fava, J.A., A.A. Jensen, L. Lindfors, S. Pomper, B, de Smet, J. Warren & B. Vigon. *Life-cycle assessment data quality. A conceptual framework.* SETAC, Pensacola, 1994.

Fecker, I. *How to calculate an ecological balance?* EMPA, St. Gallen, 1992.

Forsythe, G.E. & C.B. Moler. *Computer solution of linear algebraic systems.* Prentice Hall, Englewood Cliffs, 1967.

Freire, F., S. Thore & P. Ferrão. Life cycle activity analysis. Logistics and environmental policies for bottled water in Portugal. *OR Spektrum* 23 (2001), 159–182.

Frischknecht, R., P. Hofstetter, I. Knoepfel, E. Walder, R. Dones & E. Zollinger. *Ökoinventare für Energiesysteme. Grundlagen für den ökologischen Vergleich von Energiesystemen und den Einbezug von Energiesystemen in Ökobilanzen für die Schweiz.* ETH, Zürich, 1993.

Frischknecht R. & P. Kolm. Modellansatz und Algorithmus zur Berechnung von Ökobilanzen im Rahmen der Datenbank ECOINVENT. In: M. Schmidt & A. Schorb (Eds.). *Stoffstromanalysen in Ökobilanzen und Öko-audits.* Springer, Berlin, 1995, pp. 80–95.

Frischknecht R. Life cycle inventory modelling in the Swiss national database ECOINVENT 2000. Visit http://www.ecoinvent.ch/download/rf-envinf.pdf.

Fritsche, U., L. Rausch & K.-H. Simon. *Umweltwirkungsanalyse von En-*

ergiesystemen. Gesamt-Emissions-Modell Integrierter Systeme (GE-MIS). Endbericht. Öko-Institut, Darmstadt, 1991.

Gillies, D. *Philosophical theories of probability.* Routledge, London, 2001.

Golub, G.H. & C.F. van Loan. *Matrix Computations. Third Edition.* The Johns Hopkins University Press, Baltimore, 1996.

Graedel, T.E. *Streamlined life-cycle assessment.* Prentice Hall, Upper Saddle River, 1998.

Guinée, J.B., M. Gorrée, R. Heijungs , G. Huppes, R. Kleijn, A. de Koning, L. van Oers, A. Wegener Sleeswijk, S. Suh, H.A. Udo de Haes, H. de Bruijn, R. van Duin & M.A.J. Huijbregts. *Life cycle assessment. An operational guide to the ISO standards. I: LCA in perspective. IIa: Guide. IIb: Operational annex. III: Scientific background.* Kluwer Academic Publishers, Dordrecht, 2002.

Halada, S., K. Halada & K. Yokoyama. Sensitivity analysis in LCA using perturbation method. In: *Proceedings of the third international conference on ecobalance.* Tsukuba Research Center, Tsukuba, 1998, pp. 55–58.

Hamming, R.W. *Numerical methods for scientists and engineers. Second edition.* Dover Publications, Inc., New York, 1986.

Hanssen, O.J. & O.A. Asbjørnsen. Statistical properties of emission data in life cycle assessments. *Journal of Cleaner Production* 4:3/4 (1996), 149–157.

Harville, D.A. *Matrix algebra from a statistician's perspective.* Springer, New York, 1997.

Hauschild, M. & H. Wenzel. *Environmental assessment of products. Volume 2. Scientific background.* Chapman & Hall, London, 1998.

Häuslein, A. & J. Hedemann. Die Bilanzierungssoftware Umberto und mögliche Einsatzgebiete. In: M. Schmidt & A. Schorb (Eds.). *Stoffstromanalysen in Ökobilanzen und Öko-audits.* Springer, Berlin, 1995, pp. 59–78.

Hawkins, D.& H.A. Simon. Note. Some conditions of macroeconomic stability. *Econometrica* 17:3/4 (1949), 245–248.

Hays, W.L. *Statistics. Fourth edition.* Holt, Rinehart and Winston, Inc., New York, 1988.

Heijungs, R., J.B. Guinée, G. Huppes, R.M. Lankreijer, H.A. Udo de Haes, A. Wegener Sleeswijk, A.M.M. Ansems, P.G. Eggels, R. van Duin & H.P. de Goede. *Environmental life cycle assessment of products. I:*

Guide – October 1992. II: Backgrounds – October 1992. CML, Leiden, 1992.

Heijungs, R. A generic method for the identification of options for cleaner products. *Ecological Economics* 10:1 (1994), 69–81.

Heijungs, R. On the identification of key issues for further investigation in life-cycle screening. The use of mathematical tools and statistics for sensitivity analyses. *Journal of Cleaner Production* 4:3/4 (1996), 159–166.

Heijungs, R. *Economic drama and the environmental stage. Derivation of formal tools for environmental analysis and decision-support from a unified epistemological principle.* PhD thesis, Rijksuniversiteit Leiden, Leiden, 1997. Republished as Heijungs, R. *A theory of the environment and economic systems. A unified framework for ecological economic analysis and decision-support.* Edward Elgar, Cheltenham, 2001.

Heijungs, R. Towards eco-efficiency with LCA's prevention principle. An epistemological foundation of LCA using axioms. In: J.E.M. Klostermann & A. Tukker (Eds.). *Product innovation and eco-efficiency. Twenty-three industry efforts to reach the factor 4.* Kluwer Academic Publishers, Dordrecht, 1998, pp. 175–185.

Heijungs, R. & J.B. Guinée. Software as a bridge between theory and practice in life cycle assessment. *Journal of Cleaner Production* 1:3/4 (1993), 185–189.

Heijungs, R. & P. Hofstetter. Definitions of terms and symbols. In: H.A. Udo de Haes (Ed.). *Towards a coherent methodology for life cycle impact assessment.* SETAC, Brussels, 1995, pp. 31–37.

Heijungs, R. & R. Frischknecht. A special view on the nature of the allocation problem. *International Journal of Life Cycle Assessment* 3:5 (1998), 321–332.

Heijungs, R. & R. Kleijn. Numerical approaches towards life cycle interpretation. Five examples. *International Journal of Life Cycle Assessment* 6:3 (2001), 141–148.

Hendrickson, C., A. Horvath, S. Joshi & L. Lave. Economic input-output models for environmental life-cycle assessment. *Environmental Science & Technology* 13:4 (1998), 184A-191A.

Hendry, D.F. Monte Carlo experimentation in econometrics. In: Z. Griliches & M.D. Intriliger (Eds.). *Handbook of econometrics. Volume II.* Elsevier Science Publishers, New York, 1984, pp. 937–976.

Hoffman, L. B.P. Weidema, K. Christiansen & A.K. Ersbøll. Special reports No 1. Statistical analysis and uncertainties in relation to LCA. In: L.-

G. Lindfors, K. Christiansen, L. Hoffman, Y. Virtanen, V. Juntilla, A. Leskinen, O.-J. Hansen, A. Rønning, T. Ekvall & G. Finnveden. *LCA-Nordic. Technical reports No 10 and special reports No 1–2*. Nord, Copenhagen, 1995.

Hofstetter, P. *Perspectives in life cycle impact assessment. A structured approach to combine models of the technosphere, ecosphere and value-sphere*. Kluwer Academic Publishers, Dordrecht, 1998.

Hondo, H., K. Nishimura & Y. Uchiyama. *Energy requirements and CO_2 emissions in the production of goods and services. Application of an input-output table to life cycle analysis*. Socio-Economic Reserach Center, Tokyo, 1996 (in Japanese).

Howard, P.J.A. *An introduction to environmental pattern analysis*. The Parthenon Publishing Group, Carnforth, 1991.

Huele, R. & N. van den Berg. Spy plots. A method for visualising the structure of LCA data bases. *International Journal of Life Cycle Assessment* 3:2 (1998), 114–118.

Huijbregts, M.A.J. Application of uncertainty and variability in LCA. Part I: A general framework for the analysis of uncertainty and variability in life cycle assessment. *International Journal of Life Cycle Assessment* 3:5 (1998), 273–280. Part II: Dealing with parameter uncertainty and uncertainty due to choices in life cycle assessment. *International Journal of Life Cycle Assessment* 3:6 (1998), 343–351.

IFIAS. *Energy analysis workshop on methodology and conventions*. IFIAS, Stockholm, 1974.

ISO. *Environmental management. Life cycle assessment. Principles and framework*. ISO, Geneva, 1997.

ISO. *Environmental management. Life cycle assessment. Goal and scope definition and inventory analysis*. ISO, Geneva, 1998.

ISO. *Environmental management. Life cycle assessment. Life cycle impact assessment*. ISO, Geneva, 2000.

ISO. *Environmental management. Life cycle assessment. Life cycle interpretation*. ISO, Geneva, 2000.

Jennings, A. & J.J. McKeown. *Matrix computation. Second edition*. John Wiley & Sons, Chichester, 1977.

Jensen, A.A., L. Hoffman, B.T. Møller, A. Schmidt, K. Christiansen, J. Elkington & F. van Dijk. *Life cycle assessment (LCA). A guide to approaches, experiences and information resources*. EEEA, Copenhagen,

1997.

Johnson, R.A. & D.W. Wichern. *Applied multivariate statistical techniques. Third edition.* Prentice-Hall International, Inc., Englewood Cliffs, 1992.

Joshi, S. Product environmental life-cycle assessment using input-output techniques. *Journal of Industrial Ecology* 3:2/3 (2000), 95–120.

Kandelaars, P.P.A.A.H. *Economic models of material-product chains for environmental policy analysis.* Kluwer Academic Publishers, Dordrecht, 1999.

Kennedy, D.J., D.C. Montgomery & B.H. Quay. Data quality. Stochastic environmental life cycle assessment modeling. *International Journal of Life Cycle Assessment* 1:4 (1996), 199–207.

Kennedy, D.J., D.C. Montgomery, D.A. Rollier & J.B. Keats. Data quality. Assessing input data uncertainty in life cycle assessment models. *International Journal of Life Cycle Assessment* 2:4 (1997), 229–239.

Koopmans, T.C. *Activity analysis of production and allocation.* John Wiley & Sons, Inc., New York, 1951.

Konijn, P.J.A. *The make and use of commodities by industries. On the compilation of input-output data from national accounts.* Universiteit Twente, Enschede, 1994.

Krewitt, W., P. Mayerhofer, A. Trukenmüller & R. Friedrich. Application of the impact pathway analysis in the context of LCA. The long way from burden to impact. *International Journal of Life Cycle Assessment* 3:2 (1998), 86–94.

Lave, L.B., E. Cobas-Flores, C.T. Hendrickson & F.C. McMichael. Using input-output analysis to estimate economy-wide discharges. *Environmental Science & Technology* 29:9 (1995), 420A-426A.

Legendre, P. & L. Legendre. *Numerical ecology. Second English edition.* Elsevier, Amsterdam, 1998.

Lenzen, M. Errors in conventional and input-output-based life-cycle inventories. *Journal of Industrial Ecology* 4:4 (2001), 127–148.

Leontief, W. Environmental repercussions and the economic structure. An input-output approach. *Review of Economics and Statistics* LII (1970), 262–271.

Lipsey, R.G. & P.O. Steiner. *Economics. Fifth edition.* Harper & Row, New York, 1978.

Lindfors, L.-G., K. Christiansen, L. Hoffman, Y. Virtanen, V. Juntilla, O.-J. Hansen, A. Rønning, T. Ekvall & G. Finnveden. *Nordic guidelines on life-cycle assessment.* Nord, Copenhagen, 1995.

Lübkert, B., Y. Virtanen, M. Mühlberger, J. Ingman, B. Vallance & S. Alber. *Life-cycle analysis. IDEA. An international database for ecoprofile analysis. A tool for decision makers.* IIASA, Laxenburg, 1991.

Marheineke, T., R. Friedrich & W. Krewitt. *Application of a hybrid-approach to the life cycle inventory analysis of a freight transport task.* SAE Techical Paper Series 982201, Warrendale, 1998.

Maurice, B., R. Frischknecht, V. Coelho-Schwirtz & K. Hungerbühler. Uncertainty analysis in life cycle inventory. Application to the production of electricity with French coal power plants. *Journal of Cleaner Production* 8 (2000), 95–108.

Melo, M.T. *A review and critique of life cycle inventory methods.* Unpublished manuscript, 1999.

Miller, R.E. & P.D. Blair. *Input-output analysis. Foundations and extensions.* Prentice-Hall, Inc., Englewood Cliffs, 1985.

Morgan, M.G. & M. Henrion. *Uncertainty. A guide to dealing with uncertainties in quantitative risk and policy analysis.* Cambridge University Press, Cambridge, 1990.

Moriguchi, Y., Y. Kondo & H. Shimizu. Analysing the life cycle impact of cars: the case of CO_2. *UNEP Industry and Environment* January, 1993.

Möller, F.-J. *Ökobilanzen erstellen und anwenden. Entwicklung eines Untersuchungsmodells für die Umweltverträglichkeit von Verpackungen.* E-cobalance Applied Research, München, 1992.

Möller, A. & A. Rolf. Methodische Ansätze zur Erstellung von Stoffstromanalysen unter besonderer Berücksichtigung von Petri-Netzen. In: M. Schmidt & A. Schorb (Eds.). *Stoffstromanalysen in Ökobilanzen und Öko-audits.* Springer, Berlin, 1995, pp. 33–58.

Neter, J., M.H. Kutner, C.J. Nachtsheim & W. Wasserman. *Applied linear statistical models. Fourth edition.* IRWIN, Chicago, 1996.

Nielsen, A.M. & B. Pedersen Weidema (Eds.). *Input/output analysis. Shortcuts to life cycle data?* Miljøstyrelsen, Copenhagen, 2001.

Noh, J., S. Suh & K.M. Lee. Methods for the key issue identification in Life Cycle Assessment. In: *Proceedings of the third international conference on ecobalance.* Tsukuba Research Center, Tsukuba, 1998, pp. 59–62.

Perrings, C. *Economy and environment. A theoretical essay on the inter-*

dependence of economic and environmental systems. Cambridge University Press, Cambridge, 1987.

Pohl, C., M. Roš, B. Waldeck & F. Dinkel. Imprecision and uncertainty in LCA. In: S. Schaltegger (Ed.). *Life cycle assessment (LCA). Quo vadis?* Birkhäuser Verlag, Basel, 1996.

Pohl, C.E. *Auch zu präzis ist ungenau! Unsicherheitsanalyse in Ökobilanzen und alternativen zu "use many methods."* PhD thesis, Eidgenössischen Technischen Hochschule Zürich, Zürich, 1999.

Projektgemeinschaft Lebenswegbilanzen. *Umweltprofille von Packstoffen und Packmitteln. Methode (Entwurf).* ILV, München, 1991.

Press, W.H., B.P. Flannery, S.A. Teukolsky & W.T. Vetterling. *Numerical recipes in Pascal. The art of scientific computing.* Cambridge University Press, Cambridge, 1992.

Raa, T. ten, D. Chakraborty & J.A. Small. An alternative treatment of secondary products in input-output analysis. *The Review of Economics and Statistics* 66:1 (1984), 88–97.

Raa, T. ten. An alternative treatment of secondary products in input-output analysis. Frustration. *The Review of Economics and Statistics* 70:3 (1988), 535–538.

Reif, F. *Statistical Physics. Berkely Physics Course. Volume 5.* McGraw-Hill Book Company, New York, 1964.

Roš, M. *Unsicherheit und Fuzziness in ökologischen Bewertungen. Orientierung zu einer robusten Praxis der Ökobilanzierung.* PhD thesis, Eidgenössischen Technischen Hochschule Zürich, Zürich, 1998.

Schmidt, M. Die Modellierung von Stoffrekursionen in Ökobilanzen. In: M. Schmidt & A. Schorb (Eds.). *Stoffstromanalysen in Ökobilanzen und Öko-audits.* Springer, Berlin, 1995, pp. 97–117.

Schmidt, M. & A. Schorb (Eds.). *Stoffstromanalysen in Ökobilanzen und Öko-audits.* Springer, Berlin, 1995.

Schmidt, M. & A. Häuslein (Eds.). *Ökobilanzierung mit Computerunterstützung. Produktbilanzen und betriebliche Bilanzen mit dem Programm Umberto.* Springer, Berlin, 1997

Schuler, H. (Ed.). *Prozeßsimulation.* VCH, Weinheim, 1995.

Sebald, A.V. *An analysis of the sensitivity of large scale input-output models to parametric uncertainties.* Center for Advanced Computation, University of Illinois at Urbana-Champaign, Urbana, 1974.

Sheskin, D.J. *Handbook of parametric and nonparametric statistical proce-*

dures. CRC Press, Boca Raton, 1997.

Siegenthaler, C.P., S. Linder & F. Pagliari. *LCA software guide 1997. Market overview. Software portraits.* ÖBU, Adliswil, 1997.

Smith, A.E., P.B. Ryan & J.S. Evans. The effect of neglecting correlations when propagating uncertainty and estimating the population distribution of risk. *Risk Analysis* 12:4 (1992), 467–474.

Steen, B., R. Carlson& G. Löfgren. *SPINE. A relation database structure for life cycle assessments.* IVL, Göteborg, 1995.

Stewart, G.W. *Introduction to matrix computations.* Academic Press, Inc., Orlando, 1973.

Stewart, G.W. Stochastic perturbation theory. *SIAM Review* 32:4 (1990), 579–610.

Stewart, G.W. & J. Sun. *Matrix perturbation theory.* Academic Press, Inc., Boston, 1990.

Suh, S. *MIET 2.0 user's guide. An inventory estimation tool for missing flows using input-output techniques.* CML, Leiden, 2001.

Suh, S. & G. Huppes. *Gearing input-output model to LCA. Part I: General framework for hybrid approach.* CML, Leiden, 2000.

Suh, S. & G. Huppes. *Techniques for life cycle inventory of a product.* CML, Leiden, 2002.

Taylor, J.R. *An introduction to error analysis. The study of uncertainties in physical measurements.* University Science Books, Mill Valley, 1982.

Téno, J.-F. Le. Visual data analysis and decision support for non-deterministic LCA. *International Journal of Life Cycle Assessment* 4:1 (1999), 41–47.

Thisted, R.A. *Elements of statistical computing. Numerical computation.* Chapman and Hall, New York, 1988.

Treloar, G.J. Extracting embodied energy paths from input-output tables: towards an input-output-based hybrid energy analysis method. *Economic Systems Research* 9:4 (1997), 375–391.

UNEP. *Towards the global use of life cycle assessment.* UNEP, Paris, 1999.

Victor, P.A. *Pollution. Economy and Environment.* George Allen & Unwin, Ltd., London, 1972.

Vigon, B. Software systems and databases. In: M.A. Curran (Ed.) *Environmental life-cycle assessment.* McGraw-Hill, New York, 1996, p.3.1–3.25

Waugh, F.V. Inversion of the Leontief matrix by power series. *Econometrica* 18 (1950), 142–154.

Weidema, B.P. *Environmental assessment of products. A textbook on life cycle assessment.* TEK, Helsinki, 1997.

Weidema, B. Avoiding co-product allocation in life-cycle assessment. *Journal of Industrial Ecology* 4:3 (2001), 11–33.

Weidema, B.P. & M.S. Wesnæs. Data quality management for life cycle inventories. An example of using data quality indicators. *Journal of Cleaner Production* 4:3/4 (1996), 167–174.

Wenzel, H., M. Hauschild & L. Alting. *Environmental assessment of products. Volume 1. Methodology, tools and case studies in product development.* Chapman & Hall, London, 1998.

Index

Eco-Efficiency in Industry and Science

1. J.E.M. Klostermann and A. Tukker (eds.): *Product Innovation and Eco-efficiency.* Twenty-three Industry Efforts to Reach the Factor 4. 1997   ISBN 0-7923-4761-7
2. K. van Dijken, Y. Prince, T. Wolters, M. Frey, G. Mussati, P. Kalff, O. Hansen, S. Kerndrup, B. Søndergård, E. Lopes Rodrigues and S. Meredith (eds.): *Adoption of Environmental Innovations.* The Dynamics of Innovation as Interplay Between Business Competence, Environmental Orientation and Network Involvement. 1999
ISBN 0-7923-5561-X
3. M. Bartolomeo, M. Bennett, J.J. Bouma, P. Heydkamp, P. James, F. de Walle and T. Wolters: *Eco-Management Accounting.* 1999   ISBN 0-7923-5562-8
4. P.P.A.A.H. Kandelaars: *Economic Models of Material-Product Chains for Environmental Policy Analysis.* 1999   ISBN 0-7923-5794-9
5. J. de Beer: *Potential for Industrial Energy-Efficiency Improvement in the Long Term.* 2000   ISBN 0-7923-6282-9
6. K. Green, P. Groenewegen and P.S. Hofman (eds.): *Ahead of the Curve.* Cases of Innovation in Environmental Management. 2001   ISBN 0-7923-6804-5
7. J.B. Guinée (ed.): *Handbook on Life Cycle Assessment.* Operational Guide to the ISO Standards. 2002   ISBN 1-4020-0228-9
8. T.J.N.M. de Bruijn and A. Tukker (eds.): *Partnership and Leadership.* Building Alliances for a Sustainable Future. 2002   ISBN 1-4020-0431-1
9. M. Bennett, J.-J. Bouma and T. Wolters (eds.): *Environmental Management Accounting.* Informal and Institutional Developments. 2002
ISBN 1-4020-0552-0; Pb: ISBN 1-4020-0553-9
10. N. Wrisberg and H.A. Udo de Haas (eds.): *Analytical Tools for Environmental Design and Management in a Systems Perspective.* 2002   ISBN 1-4020-0626-8
11. R. Heijungs and S. Suh: *The Computational Structure of Life Cycle Assessment.* 2002
ISBN 1-4020-0672-1

KLUWER ACADEMIC PUBLISHERS – DORDRECHT / BOSTON / LONDON